LECTURES ON GROUPS AND VECTOR SPACES

For Physicists

World Scientific Lecture Notes in Physics Vol. 31

LECTURES ON GROUPS AND VECTOR SPACES

For Physicists

Chris J Isham

Theoretical Physics Department
Imperial College
England

World Scientific
Singapore • New Jersey • London • Hong Kong

Published by

World Scientific Publishing Co. Pte. Ltd.,
P O Box 128, Farrer Road, Singapore 9128
USA office: 687 Hartwell Street, Teaneck, NJ 07666
UK office: 73 Lynton Mead, Totteridge, London N20 8DH

LECTURES ON GROUPS & VECTOR SPACES FOR PHYSICISTS

ISBN 9971-50-954-7
     9971-50-955-5 (pbk)

Printed in Singapore by JBW Printers & Binders Pte. Ltd.

PREFACE

The theory of groups and vector spaces has many important applications in a number of branches of modern theoretical physics. These include the formal theory of classical mechanics, special and general relativity, solid state physics, general quantum theory, and modern elementary particle physics. For this reason, the Theoretical Physics Option of the final year BSc. physics programme at Imperial College has for many years included a course entitled "Algebra and Groups" and these notes are the contents of the course as given by the author during the last few years.

Naturally, one aim of the course is to provide an introduction to the general theory of groups and vector spaces, with particular emphasis being placed on the potential applications of the mathematical ideas to the structural foundations of quantum theory.

However another, and equally important, aim of the course is to introduce the student to the art of genuine 'mathematical' thinking. The physics students at Imperial College attend several ancillary mathematics courses as part of their undergraduate studies, but these are almost entirely concerned with so-called 'mathematical methods' (special functions, complex analysis, matrices, three-dimensional vector calculus etc.) and, being taken by the entire undergraduate class, are presented in a way that tends to minimize the underlying mathematical structure *per se*. The 'Algebra and Groups' course is intended to partially remedy this situation for those students engaged in the Theory Option, and this is reflected in the notes which are written in a more precise mathematical style than is often the case in courses aimed at physics students. Quite apart from the general educational value of such an exposure to abstract thinking, it is also the case that much modern theoretical physics draws on sophisticated ideas from pure mathematics. It therefore seems most important that a prospective graduate student can approach these subjects without experiencing a total shock at a too sudden exposure to such an alien culture as pure mathematics!

The course is divided into three parts. The first is a short introduction to general group theory, with particular emphasis being placed on the matrix Lie groups that play such a crucial role in modern theoretical physics. The second part deals with the theory of vector spaces, and in particular the theory of Hilbert spaces and the analytical techniques that are needed to handle the infinite-dimensional situation. The last part of the course is a short introduction to the theory of group representations and the associated theory of characters.

Finally, let me emphasise that what is presented here are just the *notes* as handed out to the students. In the context of the course, they are naturally reinforced by my verbal comments. I must apologize if the reader finds that the absence of these expository remarks renders the written material rather pithy.

CONTENTS

CONTENTS

vi

1. GROUPS

1.1. MONOIDS

A group is a mathematical set equipped with a law for combining any two elements to produce a third element in the set. This law of combination is required to satisfy certain crucial axioms which, over the years, have been found to produce a mathematical structure of exceptional importance and interest. One of the most significant concrete examples of group structure is associated with the family of maps of any set into itself. This specific example is in many ways paradigmatic for the entire theory of groups, and it is with this that we shall begin the Course.

Let X and Y be any pair of sets. A *map* (or *function*) from X to Y is an assignment of a unique element of Y to each element of X. If f is such a function then we often write $f: X \to Y$ to indicate the pair of spaces X and Y involved as well as the map itself. The unique element in Y associated with a specific element x in X is denoted by $f(x)$ (or sometimes by f_x). We shall frequently deal with the family of all possible maps from X to Y and this set will be denoted Map (X, Y).

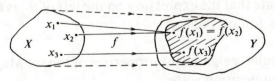

1

Note that a typical map from X to Y will be:

(a) Many–to–one: Thus more than one element in X is taken to the same element in Y. In the above diagram, we have $f(x_1) = f(x_2)$ so that x_1 and x_2 fall into this category;

(b) Strictly "into" Y: There will, in general, be element y in Y for which there is *no* element x in X for which $y = f(x)$.

     A map from X to Y which is both one–to–one and "onto" is called a *bijection* from X to Y. It establishes a unique correspondence between the elements of X and Y which means that, from a purely set theoretic viewpoint, these can be regarded as the 'same' set. Bijections between X and Y will exist if and only if they have the same 'number' of elements.

     For our purposes, the set Map(X, Y) of all maps of a set X into itself is of special importance. This is because given $f : X \rightarrow X$ and $g : X \rightarrow X$, we can combine them to form a third function $f {\circ} g : X \rightarrow X$ which is defined on any element x in X by first mapping it to $g(x)$ and then mapping this image point to $f(g(x))$. This is illustrated in the following diagram.

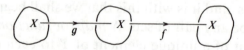

Thus $f {\circ} g : X \rightarrow X$ is defined by

$$f {\circ} g(x) := f(g(x)) \text{ for all } x \text{ in } X. \tag{1.1}$$

Note. I have introduced here the useful symbol $:=$. This is used to indicate that the expression on the left of $:=$ is to be *defined* by the expression on the right.

     Two particularly significant properties of the set Map(X, X) and the law of composition $\circ$ are:

(a) If f, g, h are three maps from X to X then

$$f \circ (g \circ h) = (f \circ g) \circ h ; \qquad (1.2)$$

(b) The *identity* map from X onto itself is denoted id_X and is defined in the obvious way as

$$id_X(x) := x \quad \text{for all } x \text{ in } X. \qquad (1.3)$$

Then it follows that, for any function $f : X \to X$, we have

$$f \circ id_X = id_X \circ f = f. \qquad (1.4)$$

The key features of this particular example $\text{Map}(X, X)$ are now abstracted out and placed in the following formal definition.

Definitions.

(a) A *law of composition* on a set A is a rule that associates with each pair of elements (a_1, a_2) (where a_1 and a_2 belong to A), a third elements of A is written as $a_1 a_2$.

(b) The law is *associative* if

$$a_1(a_2\, a_3) = (a_1\, a_2)\, a_3 \qquad \text{for any } a_1, a_2 \text{ and } a_3 \text{ in } A. \quad (1.5)$$

(c) An element e in A is said to be a *unit element* if

$$ae = ea = a \qquad \text{for all } a \text{ in } A. \qquad (1.6)$$

(d) A set A is a *monoid* if it has a law of composition that is associative and for which there is a unit element.
(We shall see shortly that the idea of a monoid is natural precursor to the concept of a group.)

Note. If such a unit e exists then it is unique. For if e' is any other unit, we have $e = e\, e' = e'$.

Examples.

(1) The set of mappings $\text{Map}(X, X)$ is a monoid for any set X. The law of composition for $f : X \to X$ and $g : X \to X$ is defined as in Eq. (1.1), i.e., $fg := f \circ g$. The unit element is simply $e := id_X$, and then Eqs. (1.2), (1.4) show precisely that we have a monoid structure on the set of maps $\text{Map}(X, X)$. This is one of the paradigmatic examples of a monoid.

(2) The set of integers $\mathbb{Z}$ is a monoid if the law of composition is ordinary addition with the unit element being the number 0.

(3) The set of integers also has a monoid structure in which the law of composition is defined to be ordinary multiplication. In this case, the unit element is the number 1.

Generally speaking, the 'product' $a_1 a_2$ of a pair of elements a_1 and a_2 in a monoid will not be the same element as the product $a_2 a_1$. However, the monoids for which all such products *are* equal are of particular importance and warrant a special definition.

Definition.

A monoid A is said to be *commutative* (or *abelian*) if

$$a_1 a_2 = a_2 a_1 \text{ for all } a_1 \text{ and } a_2 \text{ in } A.$$

In this case, it is conventional to write the product $a_1 a_2$ as $a_1 + a_2$ and to write the unit element as 0.

Examples.

(1) The set of integers is an abelian monoid with regard to both monoid structures defined in the above example.

(2) In general, the monoid $\text{Map}(X, X)$ is not commutative. For

example, let $\mathbb{R}$ denote the real numbers. Consider $\text{Map}(\mathbb{R}, \mathbb{R})$ and the two particular functions

$$f:\mathbb{R} \to \mathbb{R} \qquad f(x) := x^2$$
$$g:\mathbb{R} \to \mathbb{R} \qquad g(x) := x + 1 \, ,$$

then $f \circ g(x) = (x + 1)^2$ whereas $g \circ f(x) = x^2 + 1$, and of course these are not the same.

(3) An important example of a monoid is provided by the set $M(n, \mathbb{R})$ of all $n \times n$ real matrices. The composition of two matrices M_1 and M_2 is defined to be conventional matrix multiplication:

$$(M_1 M_2)_{ij} := \sum_{k=1}^{n} M_{1ik} M_{2kj}$$

and the unit element is the unit matrix $\mathbf{1} := \text{diag}(1, 1, \ldots, 1)$. This monoid structure is clearly non-abelian and the same applies to the analogous monoid structure defined on the set $M(n, \mathbb{C})$ of all $n \times n$ complex matrices.

(4) However, as in the case of the integers, there is another monoid structure that can be defined on $M(n, \mathbb{R})$ (and similarly on $M(n, \mathbb{C})$) which *is* abelian. This involves defining the composition of two matrices as the *sum* of the matrices, rather than the product. In this case, the unit element is the null matrix, i.e., the matrix whose elements are all zero.

1.2. THE BASIC IDEA OF A GROUP

The essential difference between a monoid and a group is that, in the latter, every element has an *inverse*. This property is of fundamental significance for the applicability of group theory to physics, and is formalized in the following definition.

Definitions.

(a) An element b in a monoid A is said to be an *inverse* of an element a in A if

$$ba = ab = e. \qquad (2.1)$$

(b) A *group* is a monoid in which every element has an inverse.

Note. If b and b' are both inverses of a then they are equal since

$$b' = b'e = b'(ab) = (b'a)b = eb = b.$$

Thus inverses are unique and it is meaningful therefore to speak of *the* inverse of an element a in A. The inverse of an element a is usually written as a^{-1}.

Examples.

(1) The set of integers $\mathbb{Z}$ is an abelian group with respect to the monoid structure in which composition is defined as addition. The inverse of an integer n is clearly $-n$.
This set $\mathbb{Z}$, however, is *not* a group under the alternative monoid structure in which composition is defined as multiplication; the inverse of an integer n would have to be $1/n$, but this is not itself an integer!

(2) The set $\mathbb{Q}$ of rational numbers is an abelian group under the law of addition.

(3) The set $\mathbb{Q}_*$ of non-zero rational numbers is an abelian group under multiplication if the inverse of the rational number n/m is defined to be the rational number m/n.

(4) In the monoid $\mathrm{Map}(X, X)$, the inverse of a function $f: X \to X$ would be a function $g: X \to X$ such that $f \circ g = g \circ f = id_X$, i.e.,

$$f(g(x)) = g(f(x)) = x \text{ for all } x \text{ in } X. \tag{2.2}$$

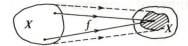

A function f will not have an inverse if it is many-to-one or if it maps strictly 'into' X. Such functions will always exist whenever the set X contains more than one element. Hence, except in that rather trivial case, Map(X, X) is never a group.

This property of Map(X, X) motivates the following, very important, definition.

Definitions.

(a) A map $f: X \to X$ is said to be a *bijection* (or *permutation*) of X if
   (i) f is a one-to-one map (i.e., it is *injective*)
   (ii) f maps X onto itself (i.e., it is *surjective*).

(b) Every such map has an inverse and it follows that the set Perm(X) of all bijections of X onto itself is a group. If X has $N < \infty$ elements then Perm (X) is often called the *symmetric group* S_N.

Note. The number of elements in a group G is called the *order* of the group and is written as $|G|$. It is easy to see (Exercise) that the order of S_N is $N!$.

When manipulating groups, it is convenient to define the 'powers' of a group element g in G by $g^2 := gg$, $g^3 := g(g^2)$ and so on. We then say that an element g has *order* $n < \infty$ if $g^n = e$ and n is the smallest integer with this property.

We must now discuss a number of simple examples of groups. The question of classifying all groups is, to put it mildly, somewhat intractable but a natural starting point is to consider groups whose order is finite and to try and classify the groups with any given value for the order. (Note that the groups $\mathbb{Z}$, $\mathbb{Q}$ and $\mathbb{Q}_*$ mentioned above are all examples of infinite-order groups since they each have an infinite number of elements.)

Order 1 groups

Clearly there is just one such group and its only element must be the unit element (which every group must possess). Thus the only product in the group is the product e^2 of the unit with itself, and of course $e^2 = e$.

Order 2 groups

Let us denote the group element that is not the unit by a. Thus $G = \{e, a\}$, where we have introduced the conventional notation that the elements of a set are labelled by the symbols appearing between the brackets { and }. Specifying the group structure on G means listing the products of all pairs of elements in G; in the present example, this entails finding all possible values for the four products ee, ea, ae, aa. We know at once that the first three must be e, a, and a, respectively because of the defining property of the unit element e. Thus the only question concerns the product a^2.

A priori there can only be two possibilities: $a^2 = a$ and $a^2 = e$. Hence there can be at most two different group structures for a group of order 2.

However, suppose that $a^2 = a$. Then,

$$a = ae = a(a \cdot a^{-1}) = (a^2) \cdot a^{-1} \text{ (by associativity)}$$
$$= a \cdot a^{-1} = e.$$

Note. The only purpose of the 'decimal point' in the expressions above is the typographical one of making it clear that the two elements on each side of the point are being combined together under the group law.

Thus $a = e$, which is a contradiction since we are assuming that the group is of order 2. Hence, the only allowed value for the product a^2 is $a^2 = e$. It is a straightforward exercise to show that this product law satisfies all the axioms for a group.

The group law for this unique group of order 2 is summarized in the 'group table' below in which the entry corresponding to a particular row and column is to be read as the product $g_1 g_2$ of the row element g_1 and the column element g_2.

$$
\begin{array}{c|cc}
 & e & a \\
\hline
e & e & a \\
a & a & e
\end{array}
$$

This unique group of order 2 is known as $\mathbb{Z}_2$ – a notation whose significance and motivation will appear later.

Note that, as it stands, this definition of the group $\mathbb{Z}_2$ is a purely abstract one expressed in terms of the combination law of the group elements. However, the question naturally arises of the existence of 'concrete' examples of this group, by which I mean a set of specific mathematical objects with a specific group combination law which reproduces this abstract structure of the group $\mathbb{Z}_2$. There are in fact many such examples, and here are a few:

(1) Let $X := \{1, 2\}$, i.e., a set with just two elements which, for convenience, I have labelled '1' and '2'. Then $\mathrm{Perm}(X)$ is a concrete group of bijections and is of order 2. The two group elements are the identity map (which is the group unit) and the map which takes '1' into '2' and '2' into '1'. If we indicate these maps with the symbols:

$$
\begin{bmatrix} 1 \to 1 \\ 2 \to 2 \end{bmatrix} \quad \text{and} \quad \begin{bmatrix} 1 \to 2 \\ 2 \to 1 \end{bmatrix}
$$

respectively, then it is clear that

$$\begin{bmatrix} 1 \to 2 \\ 2 \to 1 \end{bmatrix} \begin{bmatrix} 1 \to 2 \\ 2 \to 1 \end{bmatrix} = \begin{bmatrix} 1 \to 1 \\ 2 \to 2 \end{bmatrix}$$

which does indeed reproduce the group law of $\mathbb{Z}_2$. In this sense, Perm $\{1, 2\} = S_2$ is a "concrete example" of the abstract group $\mathbb{Z}_2$.

(2) Let $G := \{1, -1\}$ with the law of composition on the numbers 1 and -1 being multiplication. This clearly equips G with a group structure in which the unit element is the number 1. Since $(-1)^2 = 1$ we again have a concrete manifestation of $\mathbb{Z}_2$.

(3) In the x–y plane, let e denotes the identity transformation and let r be the transformation $(x, y) \to (x, -y)$, i.e., reflection in the x-axis. The set $\{e, r\}$ is a subset of the group of bijections of the plane and is itself a group. Since $r^2 = e$, this provides another example of a concrete copy of the abstract group $\mathbb{Z}_2$.

Thus, we see that the three groups Perm $\{1, 2\}, \{1, -1\}$ and $\{e, r\}$ all have the same group law as $\mathbb{Z}_2$ which, given that they *are* groups, is hardly surprising since we have already shown above that there is only one possible group structure for a group with two elements. In this sense, we could affirm that they are all the same group $\mathbb{Z}_2$. However, the actual mathematical objects in the three sets are different, and in this sense they are *not* the same group. What is needed to avoid this potentially confusing situation is a precise definition of when two concretely different mathematical sets equipped with a group structure can be regarded as the same abstract group. This is provided by the following concept of "isomorphic" groups.

Definition.

Two groups G_1 and G_2 are said to be *isomorphic* (written as $G_1 \cong G_2$) if their elements can be put in a one-to-one correspondence in a way that preserves the group combination law.

More precisely, there must exist a bijection $i: G_1 \to G_2$ such that

$$i(ab) = i(a) \cdot i(b) \quad \text{for all } a \text{ and } b \text{ in } G_1. \tag{2.3}$$

The map i is said to be an *isomorphism* of G_1 with G_2.

Note. (1) On the left hand side of Eq. (2.3), the symbol 'ab' refers to the combination according to the group law of G_1 whereas the product on the right hand side is with respect to the group law of G_2.

(2) It follows automatically from Eq. (2.3) that $i(e_1) = e_2$, where e_1 and e_2 are the unit elements in G_1 and G_2, respectively.

(3) An isomorphism of a group with itself is known as an *automorphism* of the group. The set of all of these is denoted by $\text{Aut}(G)$ and is itself a group. (Exercise: Prove this.)

Example.

Let $G_1 := \{1, -1\}$ with the group law as defined above, and let G_2 be the permutation group $\text{Perm}\{1, 2\}$ on the set $\{1, 2\}$. Then an isomorphism of G_1 with G_2 is given by the map:

$$i(1) := \begin{bmatrix} 1 \to 1 \\ 2 \to 2 \end{bmatrix}; \quad i(-1) := \begin{bmatrix} 1 \to 2 \\ 2 \to 1 \end{bmatrix}.$$

Thus the correct way of expressing the relation between the two groups $\{1, -1\}$ and $\text{Perm}\{1, 2\}$ is to say that they are isomorphic copies of the abstract group $\mathbb{Z}_2$.

Order 3 groups

Let us denote the three elements in the group by e, a and b

where, as usual, e is the unit element. The only part of the group table that we know *a priori* is

	e	a	b
e	e	a	b
a	a	?	?
b	b	?	?

and our task is to fill in the "question marks" and hence find the set of all possible group structures of order 3. We will do this in a series of steps in which the various possibilities will be whittled down:

(a) The product ab can only be a, b or e. Suppose that $ab = b$. Then multiplying both sides of this equation from the right by the inverse of b gives,

$$(ab)b^{-1} = b \cdot b^{-1} \text{, i.e., } a(b \cdot b^{-1}) = e$$

which implies that $a = e$. But this is impossible in a group of order 3 and hence $ab \neq b$.

Similarly, the assumption $ab = a$ implies that $b = e$, which is also impossible. Hence, we must have the third alternative of $ab = e$.

(b) The same line of argument applied to the product ba shows that $ba = e$.

(c) The product a^2 can only be e, a, or b. Suppose that $a^2 = a$. Then multiplying by the inverse of a yields $a = e$, which is impossible. Hence $a^2 \neq a$.

Now suppose that $a^2 = e$. Then $a^2 b = b$ and hence $a(ab) = b$ which implies $a = b$ since we have just shown that $ab = e$. But

$a = b$ is impossible since the group has order 3 and hence the third possibility must hold, i.e., $a^2 = b$.

(d) The last product we need is $b^2 = a^4$. But $a^4 = a(a^3)$ and $a^3 = a(a^2) = ab$ [from (c)] $= e$[from (a)]. Hence $b^2 = a$.

Thus we see there is at most one group structure of order 3, with the group table:

	e	a	b
e	e	a	b
a	a	b	e
b	b	e	a

Exercise.

Show that this is the table for a group, i.e., the combination law is associative, and every element has an inverse.

This unique group of order 3 is abelian (i.e., viewed as a 'matrix', the group table is symmetric) and is known as $\mathbb{Z}_3$. It can be conveniently written in the form $\mathbb{Z}_3 = \{e, a, a^2\}$ with the 'generating' element a subject to the relation $a^3 = e$. This suggests the immediate generalization:

Definition.

The *cyclic group* $\mathbb{Z}_n$ is the group of order n whose elements are written as

$$\mathbb{Z}_n = \{e, a, a^2, a^3, \ldots, a^{n-1}\} \quad \text{with } a^n = e.$$

Note that this group is abelian and that the product of any pair of elements $a^p a^q$ is uniquely specified by the single relation $a^n = e$.

In fact, defining for convenience, $a^0 := e$, we have the group law

$$a^p a^q = a^{(p+q)\bmod n} \text{ with } 0 \le p < n \text{ and } 0 \le q < n.$$

In particular, the inverse of a^p is $(a^p)^{-1} = a^{n-p}$.

Concrete examples of $\mathbb{Z}_n$ may be found readily:

(1) The set of complex numbers $\{1, e^{2i\pi/n}, e^{4i\pi/n}, \ldots, e^{2i(n-1)\pi/n}\}$ forms a group under ordinary multiplication and it is clear that the group structure is that of $\mathbb{Z}_n$.

(2) Let e and a denote, respectively, the identity transformation of the x–y plane and the transformation that rotates all vectors in the plane by $2\pi/n$ radians. Then $\{e, a, a^2, \ldots, a^{n-1}\}$ is a subset of the permutation group of the plane and is isomorphic to $\mathbb{Z}_n$.

Order 4 groups

In order to discuss these groups it is necessary to first introduce the idea of the 'product' of two groups.

Definition.

The *Cartesian product* (or *outer product*, or simply *product*) of a pair of groups G_1 and G_2 is the set $G_1 \times G_2$ of all pairs (g_1, g_2) with g_1 in G_1 and g_2 in G_2, with the composition law

$$(g_1, g_2)(g_1', g_2') := (g_1 g_1', g_2 g_2') \tag{2.4}$$

and the unit element,

$$e_{G_1 \times G_2} := (e_1, e_2),$$

where e_1 and e_2 are the unit elements in G_1 and G_2, respectively.

Exercise.

Show that this does indeed define a group structure and that the order of $G_1 \times G_2$ is $|G_1 \times G_2| = |G_1| |G_2|$.

It should be noted that, in general, it is a lengthy and complicated task to show that any particular combination law on a set satisfies the axioms of a group; for example, the number of checks of associativity increases rapidly with the size of the set. It follows, therefore, that techniques such as the above, in which a genuinely new group is constructed from existing ones, are of considerable value.

This last result is of immediate relevance for our search for groups of order 4. For we know that $\mathbb{Z}_2$ has order 2 and hence it follows that the product group $\mathbb{Z}_2 \times \mathbb{Z}_2$ has order $2^2 = 4$ as required. If we write the $\mathbb{Z}_2$ group as $\mathbb{Z}_2 = \{e, \mu\}$, where $\mu^2 = e$, then the group table for $\mathbb{Z}_2 \times \mathbb{Z}_2$ can be constructed immediately with the aid of the definition in Eq. (2.4).

	(e, e)	(e, μ)	(μ, e)	(μ, μ)
(e, e)	(e, e)	(e, μ)	(μ, e)	(μ, μ)
(e, μ)	(e, μ)	(e, e)	(μ, μ)	(μ, e)
(μ, e)	(μ, e)	(μ, μ)	(e, e)	(ϵ, μ)
(μ, μ)	(μ, μ)	(μ, e)	(e, μ)	(e, e)

$$(2.5)$$

The table is rather clumsy in this form and it is convenient to define $a := (e, \mu)$, $b := (\mu, e)$, $c := (\mu, \mu)$ and $e := (e, e)$. (Hopefully, the use of the letter "e" as the unit of both $\mathbb{Z}_2 \times \mathbb{Z}_2$ and $\mathbb{Z}_2$ will not cause any confusion!) In terms of these symbols, the group table (2.5) can be rewritten in the neater form

	e	a	b	c
e	e	a	b	c
a	a	e	c	b
b	b	c	e	a
c	c	b	a	e

i.e., $a^2 = b^2 = c^2 = e$,   (2.6)
$$ab = ba = c,$$
$$bc = cb = a,$$
$$ca = ac = b.$$

This table is often found in books on group theory and the group is frequently referred to as the *four-group* V_4. Note that it is abelian.

Returning now to our search for all groups of order 4, it is clear that another example is provided by the cyclic group $\mathbb{Z}_4 := \{e, a, a^2, a^3\}$ with the relation $a^4 = e$. It is easy to see that $\mathbb{Z}_2 \times \mathbb{Z}_2$ and $\mathbb{Z}_4$ are different groups (i.e., they are not isomorphic) but what is not so obvious is the fact that these are the *only* two groups of order 4. The pleasure of proving this is left to one of the questions in the problem behind.

We could go on now to consider groups of orders 5, 6, ... but this is a somewhat endless job and gets rather tedious rapidly. Instead, we will turn to a new concept that is of central importance in both the theory and the application of group structure.

If H is a subset of a group G then any pair of elements h_1, h_2 in H can be "multiplied" using the law of composition of the group G. However, there is no general reason why the product $h_1 h_2$ should itself lie in the subset H of G: it could well lie in the complement of H in G. But in certain special cases, it happens that $h_1 h_2$ *is* in H for every pair of elements h_1 and h_2 in H. If it is also true that the unit element e belongs to H, and that if h belongs to H then so does its inverse, then the subset H is itself a group with a group structure that is "inherited" from that of G. This rather crucial concept is formalized in the following definition.

Definition.

A subset H of a group G is said to be a *subgroup* of G if

(a) The identity element e of G belongs to H.

(b) If h_1 and h_2 are in H then so is the product $h_1 h_2$.
     (We say that H is "closed" under the composition law of G.)

(c) If h belongs to H then so does its inverse h^{-1}.
     (We say that H is "closed" under inversion.)

Examples.

(1) The subset $\{e\}$ is a subgroup of any group G, and G is a subgroup of itself. These are rather trivial examples of subgroups and are often called *improper*. All other subgroups are said to be *proper*.

(2) Any proper subgroup of a finite-order group G must have an order that is greater than one but less than the order of G. In particular, any proper subgroups of the order 3 group $\mathbb{Z}_3$ must have order 2, and there is only one possibility viz. $\mathbb{Z}_2$.

Thus if $\mathbb{Z}_3 = \{e, a, a^2\}$ with $a^3 = e$ then any proper subgroup H must be of the form $H = \{e, a\}$ or $H = \{e, a^2\}$. However, $a^2 \neq e$ in $\mathbb{Z}_3$ and hence $\{e, a\}$ cannot be a $\mathbb{Z}_2$ subgroup. Similarly,

$$(a^2)^2 = a^4 = a^3 \cdot a = a$$

since $a^3 = e$ in $\mathbb{Z}_3$. Hence, $(a^2)^2 \neq e$ and so $\{e, a^2\}$ is also not a $\mathbb{Z}_2$ subgroup. But these were the only possibilities and so we conclude that $\mathbb{Z}_3$ has *no* proper subgroups at all.

(3) If $\mathbb{Z}_4 = \{e, a, a^2, a^3\}$ wth $a^4 = e$ then $(a^2)^2 = e$ and hence the subset $\{e, a^2\}$ is a proper $\mathbb{Z}_2$ subgroup of $\mathbb{Z}_4$.

Exercise: Are there any other proper subgroups?

(4) The infinite order group of integers $\mathbb{Z}$ (under the law of addition) is a proper subgroup of the infinite order group $\mathbb{Q}$ of rational numbers.

(5) If Y is a subset of the set X then $\mathrm{Perm}(Y)$ is isomorphic to a subgroup of $\mathrm{Perm}(X)$. Specifically, we define the map $j : \mathrm{Perm}(Y) \to \mathrm{Perm}(X)$ by specifying the map $j(f) : X \to X$ for any f in $\mathrm{Perm}(Y)$ as:

$$[j(f)](x) := f(x) \qquad \text{if } x \text{ is in } Y \subset X, \qquad (2.7)$$

$$:= x \qquad \text{otherwise.}$$

Exercise.

Confirm that j is indeed an isomorphism of $\mathrm{Perm}(Y)$ onto a subgroup of $\mathrm{Perm}(X)$. This involves showing that

(a) $j(f)$ is a bijection of X when f is a bijection of Y.

(b) $j(f_1 {\circ} f_2) = [j(f_1)]{\circ}[j(f_2)]$ for all functions f_1 and f_2 in $\mathrm{Perm}(Y)$.

(c) Regarded as a function of f in $\mathrm{Perm}(Y)$, $f \rightsquigarrow j(f)$ is one-to-one.

This example sounds more complicated than it really is. All it is saying is that the permutation of X induced by a permutation f of Y is defined to be f itself on those elements of X that lie in the subset Y, and that it does nothing to the elements that lie outside Y.

For example, in the special case when $X := \{1, 2, 3\}$ and $Y := \{1, 2\}$, we have $\mathrm{Perm}(X) = S_3$ and $\mathrm{Perm}(Y) = S_2$, and the isomorphism $j : S_2 \to S_3$ is

$$j\begin{bmatrix} 1 \to 1 \\ 2 \to 2 \end{bmatrix} := \begin{bmatrix} 1 \to 1 \\ 2 \to 2 \\ 3 \to 3 \end{bmatrix} \quad \text{and} \quad j\begin{bmatrix} 1 \to 2 \\ 2 \to 1 \end{bmatrix} := \begin{bmatrix} 1 \to 2 \\ 2 \to 1 \\ 3 \to 3 \end{bmatrix}. \quad (2.8)$$

Note. Strictly speaking, the example (5) above does not say that $\mathrm{Perm}(Y)$ is a subgroup of $\mathrm{Perm}(X)$ but rather that it is isomorphic to one. However, since we are regarding isomorphic groups as being equal as abstract groups, we shall not make any distinction between these two concepts unless it clarifies some statement or other.

At the very beginning of this course, it was stated that the concrete examples of groups of the form $\mathrm{Perm}(X)$ for some set X were of particular importance. It is certainly true that specific groups of this type play a fundamental role in much modern theoretical physics. But they also have a special place in the mathematical theory of groups, and that is partly because of the following famous theorem which states that *every* group can be realized as a subgroup of a group Perm(X) for some set X.

Theorem (Cayley).

Any group G is isomorphic to a subgroup of Perm(X) for some choice of the set X.

Proof.

(a) In order to prove this theorem we need to show that, for some set X, there is a map $j: G \rightarrow$ Perm(X) such that:
   (i) j is a one-to-one map of G into Perm(X),
   (ii) for all g_1, g_2 in G, $j(g_1 g_2) = j(g_1) \circ j(g_2)$. $\qquad$ (2.9)

   We make the special choice of $X = G$ and define $j: G \rightarrow$ Perm(G) by

$$j(g) := l_g, \quad \text{where } l_g(g') := gg' . \qquad (2.10)$$

Before proving that the two conditions above are satisfied, we must first show that Eq. (2.10) does actually define an element of Perm(G), i.e., for all g in G, the map l_g is a bijection of G. Thus,
   (i) l_g is a one-to-one map of G since, if $l_g(g_1) = l_g(g_2)$, then $gg_1 = gg_2$ and so $g_1 = g_2$.
   (ii) l_g is surjective (i.e., onto) since, if g' is any element of G, we have $l_g(g^{-1}g') = g'$
and so $l_g: G \rightarrow G$ is indeed a bijection of G onto itself.

(b) Now we can prove the two desired properties of $j: G \rightarrow$ Perm(G):
   (i) If $j(g_1) = j(g_2)$ for some g_1, g_2 in G then, for every g' in G we have $j(g_1)(g') = j(g_2)(g')$ and hence $l_{g_1}(g') = l_{g_2}(g')$,

i.e., $g_1 g' = g_2 g'$. In particular, this is true for $g' = e$, and hence $g_1 = g_2$. Thus, j is a one-to-one map of G into Perm(G).

(ii) Now, we shall show that j preserves the group law. For any g_1, g_2 in G, and for all g in G,

$$j(g_1 g_2)(g) = l_{g_1 g_2}(g) = (g_1 g_2)g = g_1(g_2 g)$$

$$= l_{g_1}(l_{g_2}(g)) = l_{g_1} \circ l_{g_2}(g) = j(g_1) \circ j(g_2)(g)$$

and hence $j(g_1 g_2) = j(g_1) \circ j(g_2)$ as required.

Thus $j : G \to$ Perm(G) is an isomorphism of G onto a subgroup of Perm(G).

QED.

Note. If G is a finite group of order N then the theorem asserts that G is isomorphic to a subgroup of the symmetric group S_N.

1.3. CONTINUOUS GROUPS

The examples of groups that we have considered so far have either a finite number of elements (eg. $\mathbb{Z}_N$ and S_N) or an infinite, but countable, number of elements (eg. $\mathbb{Z}$ and $\mathbb{Q}$). However, it is clear that these are not the only possibilities and a particularly important class of groups is those that have an infinite number of elements in which the infinity concerned is 'continuous' rather than 'countable'. Groups of this type play a fundamental role in modern theoretical physics.

Examples.

(1) The set $\mathbb{R}$ of all real numbers is an abelian group in which the group composition law is the usual addition of numbers. Note that the countably infinite groups $\mathbb{Z}$ and $\mathbb{Q}$ can be regarded as subgroups of $\mathbb{R}$.

Similarly, the set $\mathbb{C}$ of complex numbers is a 'continuously infinite' abelian group under addition.

(2) The set $\mathbb{R}_+$ of positive real numbers is a continuously infinite abelian group under the law of multiplication.

(3) Let $\mathbb{R}^n$ denotes the product n times of the abelian group $\mathbb{R}$. Thus the group elements are sets of real numbers $(r_1, r_2, \ldots, r_n)$ with the law of composition

$$(r_1, r_2, \ldots, r_n) + (r'_1, r'_2, \ldots, r'_n) := (r_1 + r'_1, \ldots, r_n + r'_n)$$

and with the unit element $(0, 0, \ldots, 0)$. Note that $\mathbb{R}^n$ is an abelian group, which is why the combination law has been written as a "+". It clearly has a continuously infinite number of group elements. An analogous set of remarks apply to $\mathbb{C}^n$.

(4) The set $M(n, \mathbb{R})$ of all real $n \times n$ matrices is a continuously infinite abelian group under the law of matrix addition. Note, however, that it is *not* a group under the law of matrix multiplication since not all matrices in $M(n, \mathbb{R})$ have inverses. Similar remarks apply to the set $M(n, \mathbb{C})$ of all $n \times n$ complex matrices. Note that $M(n, \mathbb{R})$ can be viewed as a subgroup of $M(n, \mathbb{C})$.

(5) The set $U(1)$ of all complex numbers of modulus one is a continuously infinite abelian group under the usual multiplication of complex numbers.

An important refinement of the notion of 'continuously infinite' is the concept of the *real dimension* of such a group. Roughly speaking, this is defined to be the number of real numbers that are needed to specify a group element. Thus the groups $\mathbb{R}$ and $\mathbb{R}^n$ have dimension 1 and n, respectively, and the group $\mathbb{C}$ has real dimension 2 as it requires 2 real numbers to specify a single complex number. Similarly, the matrix groups $M(n, \mathbb{R})$ and $M(n, \mathbb{C})$ have dimension n^2 and $2n^2$, respectively. In fact, it is easy to see that $M(n, \mathbb{R})$ is actually isomorphic to $\mathbb{R}^{n^2}$.

Another example of a one-dimensional group is the group $\mathbb{R}_+$. Indeed, every positive real number x can be written uniquely in the form $x = e^r$ for some real number r and so this real number can be used to 'parametrize' this group.

A more subtle example of a one-dimensional group is afforded by the group $U(1)$ of all complex numbers of modulus one. Every such number can be written in the form e^{ir} but the real number r is now only determined up to an addition of an arbitrary integral multiple of 2π.

It is obvious geometrically that the group $U(1)$ has the 'topological' structure of a circle. This is certainly a one-dimensional space, but one that is topologically different from the simple one-dimensional space $\mathbb{R}$.

This raises an interesting question. We have seen that, for any n, the group $\mathbb{R}^n$ is n-dimensional, but there are many other n-dimensional spaces that differ topologically from $\mathbb{R}^n$. Can any of them be given a group structure? Such a group would be n-dimensional and yet differ topologically from the simple group $\mathbb{R}^n$ in a way that was analogous to the manner in which $U(1)$ differs from $\mathbb{R}$. It should be noted that there is more to this question than meets the eye at a first glance since we do not merely require that the set of all elements in the group constitute an n-dimensional space, but also that the coordinates of the point gg' (i.e., the set of n real numbers that serve to specify the group elements in some region surrounding the point gg') should be suitably continuous (or even differentiable) functions of the coordinates of the group elements g and g'.

For example, in the case of $\mathbb{R}^n$, the coordinates of a group element $g = (r_1, r_2, \ldots, r_n)$ can be chosen to be simply the set of n real numbers $r_1, \ldots, r_n$ and then clearly the coordinates of $gg' = (r_1 + r_1', \ldots, r_n + r_n')$ are differentiable functions of the coordinates of g and g'. The imposition of such continuity/differentiable conditions is necessary for the theory of 'continuously infinite' groups to have any real content.

Examples.

(1) A simple two-dimensional space is the 2-torus, i.e., the surface of an 'American doughnut'. This is the group space of the product group $U(1) \times U(1)$ since a 2-torus is the Cartesian product of two circles, i.e., a point on the torus can be specified by the coordinates of the two equatorial circles.

(2) Another simple 2-dimensional space is the 2-sphere which can be defined as the set of all points (x, y, z) in ordinary Cartesian 3-space subject to the constraint $x^2 + y^2 + z^2$ is equal to some constant. It is not obvious, but in fact there is *no* continuously infinite group which has this 2-sphere as its group space.

(3) In higher dimensions, the n-torus can always be given a group structure since it is the group space of the Cartesian product group $U(1) \times U(1) \times \ldots \times U(1)$ n-times.

Spheres also exist in every dimension. In fact, the n-sphere is defined to be the set of all points $(x_1, x_2, \ldots, x_{n+1})$ in Cartesian $n + 1$ space subject to the constraint

$$x_1^2 + x_2^2 + \ldots + x_{n+1}^2 = \text{constant} .$$

It turns out that the only sphere that can be the group space of a continuously infinite group is the 3-sphere – an example that is actually of central importance to the whole subject.

At this stage, it would be a good idea to give a proper definition of the objects we are talking about (Lie groups) and then construct some less trivial examples than those given so far.

Definition.

A *Lie group* of real dimension n is a set G that is

(a) A group in the general sense discussed in Sec. 1.2.

(b) A n-dimensional 'differentiable manifold' in the sense that the points of G can be parametrized in sufficiently small regions by a set of *n* real numbers and that, if a second set of *n* coordinates is used to parametrize points in a region that overlaps the first then, on the overlap region, the first set of parameters/coordinates must be differentiable functions of the second set, and vice versa.

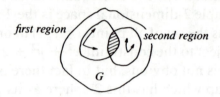

It is also required that the group composition law, and the taking of inverses, should be 'smooth' operations in the sense that

(1) The coordinates of the product gg' should be differentiable functions of the coordinates of g and g' so long as all three group elements g, g' and gg' lie in a region where a common set of coordinates can be used.

(2) The coordinates of g^{-1} should be differentiable functions of the coordinates of g as long as they lie in the same coordinate region.

Note the existence in the definition above of references to 'sufficiently small' regions and 'overlapping regions' etc. The reason for this is that on a general *n*-dimensional manifold, it is not possible to parametrize *all* of the points smoothly with a single set of *n* coordinates. Indeed, any space for which this *is* possible is necessarily topologically identical to the Cartesian *n*-space $\mathbb{R}^n$. [**Note.** When using the symbol $\mathbb{R}^n$ in this way, I am not thinking of it as a group but rather as the underlying topological space of the abelian group $\mathbb{R}^n$, i.e., the space of all *n*-tuples of real numbers $(r_1, r_2, \ldots, r_n)$.]

On a general differentiable manifold, the best that can be done is to parametrize *regions* of the manifold with sets of *n* coordinates and then make sure that on an overlap region, where two different coordinate systems are potentially in use, the different coordinates are differentiable functions of each other.

We will now give some extremely important examples of Lie groups.

Examples.

(1) The necessary and sufficient condition for a matrix in $M(n, \mathbb{R})$ to be invertible is that its determinant should be non-zero. This motivates the definition of the *general linear* group in n dimensions

$$\text{GL}(n, \mathbb{R}) := \{A \text{ in } M(n, \mathbb{R}) \text{ such that } \det(A) \neq 0\}.$$

This is a group under matrix multiplication in which the inverse of a group element is simply its inverse as a matrix in the usual sense, and the unit element is the unit matrix **1**.

A convenient set of coordinates on $\text{GL}(n, \mathbb{R})$ is provided by the n^2 matrix elements [i.e., the coordinates that it 'inherits' as a subspace of $M(n, \mathbb{R})$] and it is not too difficult to prove rigorously that $\text{GL}(n, \mathbb{R})$ is an n^2-dimensional Lie group. It should be noticed however that, unlike in the case of $M(n, \mathbb{R})$, the coordinate system provided by the matrix elements does *not* cover the whole of $\text{GL}(n, \mathbb{R})$ in a smooth way. This is because $\text{GL}(n, \mathbb{R})$ is not the same topologically as $M(n, \mathbb{R})$ (which is topologically the same as $\mathbb{R}^{n^2}$); for example, there are loops in $\text{GL}(n, \mathbb{R})$ that cannot be contracted smoothly to a point (rather like the loops going 'round' the body of a torus).

(2) Similarly, the set

$$\text{GL}(n, \mathbb{C}) := \{A \text{ in } M(n, \mathbb{C}) \text{ such that } \det(A) \neq 0\}$$

is a $2n^2$-dimensional Lie group. Note that, for $n > 1$, both $\text{GL}(n, \mathbb{C})$ and $\text{GL}(n, \mathbb{R})$ are non-abelian since, of course, for many pairs of matrices A, B, the product AB will be different from BA.

(3) The set of matrices

$$GL^+(n, \mathbb{R}) := \{A \text{ in } GL(n, \mathbb{R}) \text{ such that } \det(A) > 0\}$$

is a subgroup of $GL(n, \mathbb{R})$ since,

(i) $\det(\mathbf{1}) = 1$, i.e., the unit element $\mathbf{1}$ belongs to $GL^+(n, \mathbb{R})$,

(ii) $\det(AB) = \det(A)\det(B)$.   Hence $\det(A) > 0$ and $\det(B) > 0$ implies that $\det(AB) > 0$, i.e., AB is in $GL^+(n, \mathbb{R})$,

(iii) $\det(A^{-1}) = [\det(A)]^{-1}$. Hence, if a matrix A belongs to the subset $GL^+(n, \mathbb{R})$ then so does its inverse.

Thus $GL^+(n, \mathbb{R})$ satisfies the three conditions for being a subgroup.

Note that the imposition of the constraint $\det(A) > 0$ selects one of the two disjoint pieces into which $GL(n, \mathbb{R})$ decomposes according to the sign of $\det(A)$. Thus, like $GL(n, \mathbb{R})$, the subgroup $GL^+(n, \mathbb{R})$ is a Lie group of dimension n^2.

(4) The group of *Möbius transformations* of the complex plane $\mathbb{C}$ is defined to be the subset of $\text{Perm}(\mathbb{C})$ of transformations of the form

$$z \rightsquigarrow \frac{az + b}{cz + d} \quad \text{with } a, b, c, d \text{ in } \mathbb{C} \text{ and } ad - bc = 1. \quad (3.1)$$

This is a 6-dimensional Lie group and plays an important role in the more geometrical side of complex analysis.

Many groups of major significance in theoretical physics appear as explicit subgroups of the general linear group. For example:

Examples.

(1) The *special linear group* is defined as

$$SL(n, \mathbb{R}) := \{A \text{ in } GL(n, \mathbb{R}) \text{ such that } \det(A) = 1\}.$$

The proof that this satisfies the three conditions for being a subgroup is very similar to that given above for $GL^+(n, \mathbb{R})$. Note that, in this case, the condition $\det(A) = 1$ is a polynomial equation in the matrix elements $A_{ij}, i, j = 1, \ldots, n$, of A. It follows that $SL(n, \mathbb{R})$ is a Lie group whose dimension is one less than that of $GL(n, \mathbb{R})$, i.e., it is equal to $n^2 - 1$. (In this form, the argument is a bit heuristic, but it can be made rigorous.)

(2) Similarly,

$$SL(n, \mathbb{C}) := \{A \text{ in } GL(n, \mathbb{C}) \text{ such that } \det(A) = 1\}$$

is a Lie group of real dimension $2(n^2 - 1)$. It is a subgroup of $GL(n, \mathbb{C})$ for the same reasons that $SL(n, \mathbb{R})$ is a subgroup of $GL(n, \mathbb{R})$.

(3) The next example is of considerable importance. This is the real *orthogonal* group, which is the subgroup of $GL(n, \mathbb{R})$ of all real, $n \times n$ orthogonal matrices:

$$O(n, \mathbb{R}) := \{A \text{ in } GL(n, \mathbb{R}) \text{ such that } AA^t = \mathbf{1}\},$$

where A^t denotes the transpose of the matrix A.

This subset of $GL(n, \mathbb{R})$ satisfies the three conditions for being a subgroup since

(i) $\mathbf{1}\,\mathbf{1}^t = \mathbf{1}$ so that the unit element is in $O(n, \mathbb{R})$;

(ii) If $AA^t = \mathbf{1}$ and $BB^t = \mathbf{1}$ then $(AB)(AB)^t = (AB)B^t A^t = AA^t = \mathbf{1}$.

Thus the product of two orthogonal matrices is orthogonal;

(iii) We note that $A^{-1} = A^t$ if $AA^t = \mathbf{1}$, and hence $A^{-1}(A^{-1})^t = A^t A = A^{-1}A = \mathbf{1}$.

Thus the inverse of an orthogonal matrix is orthogonal. This subgroup $O(n, \mathbb{R})$ is also a Lie group whose dimension can be deduced by noticing that the constraint $AA^t = \mathbf{1}$ results in a number of polynomial equations on the matrix elements of the matrix A when it is regarded as lying in the set $M(n, \mathbb{R})$. Specifically,

   (i)   The diagonal elements of the matrix equation $AA^t = \mathbf{1}$ produce n polynomial equations;

  (ii) There are a further $(n^2 - n)/2$ equations coming from the matrix elements lying above the diagonal;

(iii) The matrix elements lying below the diagonal do not give any additional equations as they are identical to those obtained from the elements above the diagonal.

Thus $AA^t = \mathbf{1}$ is equivalent to $n + (n^2 - n)/2 = (n^2 + n)/2$ polynomial equations in the n^2 matrix elements of A. This implies (and this can be made rigorous) that $O(n, \mathbb{R})$ has dimension $n^2 - (n^2 + n)/2 = n(n - 1)/2$.

(4) The equation $AA^t = \mathbf{1}$ implies that $\det(A) = \pm 1$. The continuous group $O(n, \mathbb{R})$ decomposes into two disjoint pieces according to the sign of $\det(A)$, which motivates the definition of the *special orthogonal group*

$$SO(n, \mathbb{R}) := \{A \text{ in } O(n, \mathbb{R}) \text{ such that } \det(A) = 1\}.$$

The proof that this is a genuine subgroup of $O(n, \mathbb{R})$ is very similar to that showing that $GL^+(n, \mathbb{R})$ is a subgroup of $GL(n, \mathbb{R})$. The dimension of $SO(n, \mathbb{R})$ is the same as $O(n, \mathbb{R})$, namely $n(n - 1)/2$.

Note that if we denote the fact that H is a subgroup of G by the set theoretic inclusion sign $H \subset G$, then we have the following chain of subgroups

$$SO(n, \mathbb{R}) \subset O(n, \mathbb{R}) \subset GL(n, \mathbb{R}) \tag{3.2}$$
$$\subset GL^+(n, \mathbb{R}) \subset$$

but note that $O(n, \mathbb{R})$ is *not* a subgroup of $GL^+(n, \mathbb{R})$.

The simplest non-trivial example of a special orthogonal group is $SO(2, \mathbb{R})$ which is the set of all real 2×2 matrices $\begin{pmatrix} a & b \\ c & d \end{pmatrix}$ satisfying the conditions

$$\begin{pmatrix} a & b \\ c & d \end{pmatrix} \begin{pmatrix} a & c \\ b & d \end{pmatrix} = \begin{pmatrix} 1 & 0 \\ 0 & 1 \end{pmatrix} \tag{3.3}$$

$$\det \begin{pmatrix} a & b \\ c & d \end{pmatrix} = 1 . \tag{3.4}$$

Equation (3.3) is equivalent to the three equations

$$a^2 + b^2 = 1 \tag{3.5}$$

$$c^2 + d^2 = 1 \tag{3.6}$$

$$ac + bd = 0 , \tag{3.7}$$

while Eq. (3.4) implies,

$$ad - bc = 1 . \tag{3.8}$$

From Eqs. (3.7–8) we get

$$c = -\frac{bd}{a} = -\frac{b}{a} \frac{(1 + bc)}{a} \tag{3.9}$$

so that

$$c(1 + b^2/a^2) = -b/a^2 , \text{ i.e., } c = -b/(a^2 + b^2) \tag{3.10}$$

which, from Eq. (3.5), gives

$$c = -b. \tag{3.11}$$

A similar proof shows that

$$d = a. \tag{3.12}$$

Finally, Eq. (3.5) shows us that, without any loss of generality, we can write $a = \cos\theta$ and $b = \sin\theta$ for some angle θ. So the most general form for a matrix in $SO(2, \mathbb{R})$ can be written as

$$A = \begin{pmatrix} \cos\theta & \sin\theta \\ -\sin\theta & \cos\theta \end{pmatrix} \quad \text{with } 0 \le \theta < 2\pi. \tag{3.13}$$

This shows rather clearly that $SO(2, \mathbb{R})$ has the topological structure of a circle, just as does the group $U(1)$. In fact, these two abelian, 1-dimensional Lie groups are isomorphic with an isomorphism that maps the group element $e^{i\theta}$ in $U(1)$ onto the matrix in $SO(2, \mathbb{R})$ in Eq. (3.13).

The above groups were all obtained as subgroups of the real general linear group $GL(n, \mathbb{R})$. Another extremely important family of groups appears as subgroups of the complex linear group $GL(n, \mathbb{C})$.

Examples.

(1) In analogy with the real orthogonal group, the *unitary group* $U(n)$ is defined for each n as

$$U(n) := \{A \text{ in } GL(n, \mathbb{C}) \text{ such that } AA^\dagger = \mathbf{1}\},$$

where $A^\dagger$ denotes the adjoint of the matrix A and is defined as $(A^\dagger)_{ij} := A_{ji}^*$, where A_{ij}^* is the complex conjugate of the matrix element A_{ij} with $0 \le i, j \le n$. The fact that $(AB)^\dagger = B^\dagger A^\dagger$ shows that $U(n)$ is indeed a genuine subgroup of $GL(n, \mathbb{C})$ and an argument similar to the one employed above for $O(n, \mathbb{R})$ shows that the real dimension of this Lie group is n^2. (Exercise)

Note that $GL(1, \mathbb{C})$ is the same thing as the set $\mathbb{C}_*$ of all non-zero complex numbers since a 1×1 matrix is a single complex number and its determinant is just the number itself. It follows that, according to the definition just given,

$$U(1) := \{z \text{ in } \mathbb{C}_* \text{ such that } zz^* = 1\} \tag{3.14}$$

since the adjoint of a 1×1 matrix is just the complex conjugate of the corresponding complex number. But the condition (3.14) is just the statement that $U(1)$ is the set of all complex numbers of modulus one. This shows that our earlier definition of the group $U(1)$ is equivalent to the special case $n = 1$ of the general definition of $U(n)$ given in the above example. Note, however, that $U(n)$ for $n > 1$ is quite a complicated group, having dimension n^2 and also being non-abelian.

(2) A group that plays a central role in the classification of elementary particles and in the construction of grand unified theories is the *special unitary group* defined for each n as

$$SU(n) := \{A \text{ in } U(n) \text{ such that } \det(A) = 1\}.$$

By a process that should by now be familiar, it can be shown that this extra constraint does indeed define a subgroup and that the dimension of $SU(n)$ is one less than that of $U(n)$, i.e., it is equal to $n^2 - 1$.

Note that $SU(1)$ is a trivial group since it consists of just the number 1! On the other hand, the next group in the series – $SU(2)$ – is very far from being trivial, both in its mathematical significance and in its applicability to physics. To get a better handle on the structure of the 3-dimensional group SU(2), we will start by looking at the 4-dimensional group U(2). This is defined to be the set of all 2×2 complex matrices satisfying $AA^\dagger = 1$. (Note that this implies automatically that $\det(A) \neq 0$.) Thus U(2) is the set of matrices

$$\left\{ \begin{pmatrix} a & b \\ c & d \end{pmatrix} \quad \text{such that} \quad \begin{pmatrix} a & b \\ c & d \end{pmatrix} \begin{pmatrix} a^* & c^* \\ b^* & d^* \end{pmatrix} = \begin{pmatrix} 1 & 0 \\ 0 & 1 \end{pmatrix} \right\} \tag{3.15}$$

which implies that

$$|a|^2 + |b|^2 = |c|^2 + |d|^2 = 1 \tag{3.16}$$

$$ac^* + bd^* = 0 . \tag{3.17}$$

The SU(2) condition $\det(A) = 1$ imposes the additional constraint

$$ad - bc = 1 \tag{3.18}$$

on the complex parameters a, b, c, and d.

Equations (3.16–18) can be analysed in the same way that we studied Eqs. (3.5–8) for the case SO(n, $\mathbb{R}$). This leads to [cf. Eqs. (3.11–12)]

$$c = -b^* \quad \text{and} \quad d = a^* \tag{3.19}$$

so that the general SU(2) matrix is of the form

$$A = \begin{pmatrix} a & b \\ -b^* & a^* \end{pmatrix} \quad \text{with} \quad |a|^2 + |b|^2 = 1 . \tag{3.20}$$

Thus topologically, the group space of SU(2) is the set of all pairs of complex numbers (a, b) satisfying the constraint (3.20). But a complex number a corresponds to a pair of real numbers $(\text{Re}(a), \text{Im}(a))$, where $\text{Re}(a)$ and $\text{Im}(a)$ are respectively the real and imaginary parts of a. Thus we can also think of the group space of SU(2) as the set of all quadruples of real numbers $(\text{Re}(a), \text{Im}(a), \text{Re}(b), \text{Im}(b))$ satisfying the constraint

$$[\text{Re}(a)]^2 + [\text{Im}(a)]^2 + [\text{Re}(b)]^2 + [\text{Im}(b)]^2 = 1 \quad (3.21)$$

and this is just the equation of a 3-sphere embedded in the real 4-dimensional Cartesian space $\mathbb{R}^4$ with coordinates $\text{Re}(a)$, $\text{Im}(a)$, $\text{Re}(b)$, $\text{Im}(b)$, i.e.,

---

Topologically speaking, the group SU(2) is a 3-sphere.

This is a very beautiful result and is often used to illustrate the way in which geometric and group theoretic ideas are combined in the concept of a Lie group.

As a final (and physically very important) example of a Lie group, we will consider the *Weyl-Heisenberg group*. The elements in this group are the set of all collections $(\mathbf{a}, \mathbf{b}, r)$, where $\mathbf{a}$ and $\mathbf{b}$ are ordinary 3-vectors and r is a real number. Thus the group is topologically the same as the Cartesian space $\mathbb{R}^7$. However, the group law differs subtly from that of the abelian group $\mathbb{R}^7$ and is of the form

$$(\mathbf{a}, \mathbf{b}, r)\,(\mathbf{a}', \mathbf{b}', r') := \left(\mathbf{a} + \mathbf{a}', \mathbf{b} + \mathbf{b}', r + r' + \frac{1}{2}(\mathbf{b}\cdot\mathbf{a}' - \mathbf{b}'\cdot\mathbf{a})\right).$$

$$(3.22)$$

It is the presence in the last term of the additional factor $\frac{1}{2}(\mathbf{b}\cdot\mathbf{a}' - \mathbf{b}'\cdot\mathbf{a})$ that stops this just being the group law of $\mathbb{R}^7$. This extra term makes a lot of difference however and the Weyl-Heisenberg group is deeply connected with the familiar canonical commutation relations for a quantum mechanical particle moving in three dimensions

$$[\hat{q}_a, \hat{q}_b] = [\hat{p}_a, \hat{p}_b] = 0$$
$$[\hat{q}_a, \hat{p}_b] = i\,\delta_{ab} \qquad \text{for } a, b = 1, 2, 3 . \qquad (3.23)$$

1.4. GROUP OPERATIONS ON A SET

In Sec. 1.2, we saw with the aid of Cayley's theorem that any group G can be regarded as a subgroup of the group of bijections of a certain set X, namely $X = G$. This situation can be generalized greatly and, in the application of group theory to physics (and to other branches of pure mathematics), there are two questions that arise frequently and that are of considerable importance and interest

(1) Given a group G, for what sets X can G appear as one of the subgroups of Perm(X)?
(2) Conversely, given a set X, what groups G appear as subgroups of Perm(X)?

These questions can be varied in a number of useful ways but, in all cases, the basic idea is that we are looking at "representations" of a group G in the group Perm(X) of some set X. In general, to say that we have a representation of one group G_1 in another group G_2 means that there is a 'homomorphism' between the groups. This is a weakening of the notion of an isomorphism, introduced in Sec. 1.2, in which the map between the two groups is still required to preserve the group law but it is *not* required to be a bijection. The formal definition is as follows:

Definition.

A *homomorphism* between two groups G_1 and G_2 is a map $\mu : G_1 \to G_2$ which preserves the group law in the sense that

$$\mu(gg') = \mu(g)\mu(g') \quad \text{for all } g \text{ and } g' \text{ in } G. \tag{4.1}$$

Comments.

(a) Equation (4.1) implies that (Exercise)
  (i) $\mu(e_1) = e_2$, where e_1 and e_2 are the unit elements in G_1 and G_2, respectively.
  (ii) $\mu(g^{-1}) = [\mu(g)]^{-1}$ for all g in G_1.

(b) An isomorphism is simply a homomorphism that is in addition a bijection between the sets G_1 and G_2. Note that, in this case, the map $\mu^{-1}: G_2 \to G_1$ that is the inverse of the map $\mu_1: G_1 \to G_2$ is also a homomorphism (Exercise).

(c) So far as $\mu: G_1 \to G_2$ is *not* one-to-one, a 'part' of G_1 fails to be genuinely represented in G_2, and similarly, so far as μ is not surjective, a 'part' of G_2 is surplus to the needs of representing G_1. We shall return later (Sec. 1.5) to the proper formulation of these concepts.

Using the idea of a homomorphism between two groups, we can now give the definition of what it means to say that a group G is 'represented' in the group of permutations of a set X.

Definition.

A group G is said to be *represented* by the bijections of a set X, or to *act on the left* on X, if there is a homomorphism $\mu: G \to \text{Perm}(X)$.

Comments.

(a) It is common practice to write the map $\mu(g): X \to X$ as L_g, in which case the basic condition in Eq. (4.1) becomes,

$$L_{g_2} \circ L_{g_1} = L_{g_2 g_1} \quad \text{for all } g_1 \text{ and } g_2 \text{ in } G. \tag{4.2}$$

For the sake of simplicity, the point $\mu(g)(x) = L_g(x)$ in X reached by acting with g in G on the point x in X is often just written as gx, and in this notation Eqs. (4.1–2) read

$$g_2(g_1x) = (g_2g_1)x \quad \text{for all } x \text{ in } X \text{ and all } g_1, g_2 \text{ in } G.$$

(b) If the function from G into Perm(X) defined by $g \leadsto L_g$ is one-to-one then the conditions (4.2) are precisely equivalent to saying that G is isomorphic to a subgroup of Perm(X).

The other extreme case of a left G-action on X would be where $L_g := id_X$ for all g in G. In this case (which can be used to define an action of G on *any* set X), none of G is "faithfully" represented at all by Perm(X), which of course is pretty useless.

There are typically many situations between these two examples in which a group G is not mapped isomorphically into a subgroup of Perm(X), but neither is the action totally trivial. The extent of the "part" of G that *is* represented trivially is measured by a certain subgroup of G known as the *kernel* of the homomorphism $\mu : G \to \text{Perm}(X)$. This is defined to be the set of all elements of G that are mapped by μ into the identity transformation id_X: in the case of an isomorphism this is the only unit element, whereas for the trivial representation, it is the whole of G, and in the general case, it will be some proper subgroup of G.

This is an important idea and we will return to it later once we have covered the concept of a "normal" subgroup in Sec. 1.5.

(c) For a similar reason that we introduced the idea of an isomorphism between two groups, it is important to have a clear formulation of the conditions under which two G-spaces X and X' (i.e., sets on which G has a left action) can be regarded as being different copies of the "same" abstract G-space. This requires that there should be a bijection $i : X \to X'$ (so that X and X' are the "same" regarded purely as sets) which is *equivariant*, i.e., it preserves the group action in the sense that

$$i(gx) = g(i(x)) \qquad \text{for all } g \text{ in } G \text{ and } x \text{ in } X \qquad (4.3)$$

or, more precisely, the following diagram is *commutative*, which means that, for all g in G,

$$
\begin{array}{ccc}
X & \xrightarrow{\; i \;} & X' \\
\Big\downarrow{\scriptstyle L_g} & & \Big\downarrow{\scriptstyle L'_g} \\
X & \xrightarrow[\; i \;]{} & X'
\end{array}
\qquad i \circ L_g \equiv L'_g \circ i, \qquad (4.4)
$$

where L_g and L'_g are the actions on X and X'.

The full group $\mathrm{Perm}(X)$ of transformations of the set X has the property that given any two points x_1 and x_2 in X, there exists an element f in $\mathrm{Perm}(X)$ such that $x_2 = f(x_1)$. However, a subgroup of $\mathrm{Perm}(X)$, or more generally a group G acting on X, may not have this property, and this motivates the following definition.

Definitions.

(a) If G acts on a set X, the action is said to be *transitive* if any pair of points in X can be "joined" by an action of some element in G. More precisely, the action is transitive if given any pair of points x, y in X, there exists some g in G such that $gx = y$.(This relation is symmetric since $gx = y$ implies $g^{-1}y = x$, i.e., if y can be "reached" from x then x can be "reached" from y.)

(b) In general, the *orbit* of the G-action through a point x in X is defined to be the set of all points in X that can be reached by acting on x with some element of G, i.e.,

$$O_x := \{gx, \text{ where } g \text{ is in } G\}.$$

In particular, if the G-action on X is transitive then the orbit through any point x in X is the whole of X.

It should be noted that the orbits through two different points x_1 and x_2 in X must either coincide or be totally disjoint. Thus, in general, a G-action on a set X results in X being partitioned into a union of disjoint orbits.

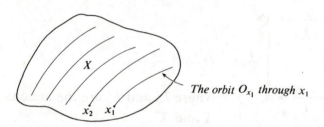

The orbit O_{x_1} through x_1

Note. The discussion so far has been in terms of a left action of a group G on a set X, but there is an equivalent notion of a *right* action. This is an association of a bijection R_g of X to each g in G such that

$$R_{g_2} \circ R_{g_1} = R_{g_1 g_2} \qquad \text{for all } g_1 \text{ and } g_2 \text{ in } G. \qquad (4.5a)$$

A right action is often written as $R_g(x) = xg$ so that condition (4.5a) reads

$$(xg_1)g_2 = x(g_1 g_2) \qquad \text{for all } g_1 \text{ and } g_2 \text{ in } G \text{ and } x \text{ in } X. \qquad (4.5b)$$

Examples.

(1) Consider the group $\mathbb{Z}_2 := \{e, a\}$ with $a^2 = e$. Then a left action of $\mathbb{Z}_2$ on the real line $\mathbb{R}$ is defined by

$$L_e(r) := r$$
$$L_a(r) := -r \qquad \qquad \text{for all } r \text{ in } \mathbb{R}.$$

The orbits of this $\mathbb{Z}_2$-action are,

$$O_r = \{r, -r\} \qquad \text{if } r \neq 0$$
$$O_0 = \{0\}.$$

(2) Let G be the group $SO(2, \mathbb{R})$, i.e., the set of all matrices

$$\begin{pmatrix} \cos\theta & -\sin\theta \\ \sin\theta & \cos\theta \end{pmatrix} \quad \text{with } 0 \leq \theta < 2\pi.$$

This group has a left action on the set of all 2×1 real column matrices given by

$$\begin{pmatrix} a \\ b \end{pmatrix} \rightsquigarrow \begin{pmatrix} \cos\theta & -\sin\theta \\ \sin\theta & \cos\theta \end{pmatrix} \begin{pmatrix} a \\ b \end{pmatrix}$$

$$= \begin{pmatrix} a\cos\theta - b\sin\theta \\ a\sin\theta + b\cos\theta \end{pmatrix}.$$

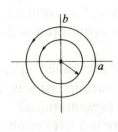

It is clear that $a^2 + b^2$ is constant on the orbits of the $SO(2, \mathbb{R})$ action and in fact the orbit of a column matrix $\begin{pmatrix} a \\ b \end{pmatrix}$ is a circle of radius $a^2 + b^2$ if $a \neq 0$ and $b \neq 0$. If $a = b = 0$, the orbit is just that single column matrix $\begin{pmatrix} 0 \\ 0 \end{pmatrix}$.

This is a very simple example of the general, and powerful, concept of a group acting on a vector space as a group of linear transformations. This idea will be developed in details in the third of the three major sections in this course.

(3) Any group G acts on itself on the left by means of left

multiplication (often known as *translation* in this context) defined as

$$L_g(g') := gg' . \tag{4.6}$$

Notice that this action is transitive (so that the orbit through any element g' in G is the whole of G) and that it was in fact what we used in the proof of Cayley's theorem in Sec. 1.2. Another transitive left action of G on itself is

$$L_g(g') := g' g^{-1} . \tag{4.7}$$

(4) Another, and very important, example of an action of a group G on itself is known as *conjugation* and is defined as

$$L_g(g') := g \, g' \, g^{-1} . \tag{4.8}$$

Notice that, unlike in the two previous cases, this particular group action of G on itself is *not* transitive. For example, since $L_g(e) = e$ for all g in G, it follows that the orbit through the unit element e is just the set $\{e\}$.

The orbits of the conjugation action are called *conjugacy classes* and they play a crucial role in various aspects of the general structure of groups and their representations.

(5) In analogy with example (2) above, there is a left action of the group $GL(n, \mathbb{R})$ on the space of all $n \times 1$ real column matrices (which is theoretically the same set as the Cartesian space $\mathbb{R}^n$) given by matrix multiplication. Thus if A is a real, $n \times n$ matrix in $GL(n, \mathbb{R})$ and if $\mathbf{r} = (r_1, r_2, \ldots, r_n)$, then the $GL(n, \mathbb{R})$ action is $\mathbf{r} \rightsquigarrow \mathbf{r}' = L_A(\mathbf{r})$ with

$$\begin{pmatrix} r'_1 \\ r'_2 \\ \vdots \\ r'_n \end{pmatrix} = \begin{pmatrix} A_{11} & A_{12} & \cdots & A_{1n} \\ A_{21} & A_{22} & \cdots & A_{2n} \\ \vdots & \vdots & \vdots\vdots\vdots & \vdots \\ A_{n1} & A_{n2} & \cdots & A_{nn} \end{pmatrix} \begin{pmatrix} r_1 \\ r_2 \\ \vdots \\ r_n \end{pmatrix},$$

i.e., $r'_i := \sum_{j=1}^{n} A_{ij} r_j$ for $i = 1, \ldots, n$. \hfill (4.9)

However, as we know, $\mathbb{R}^n$ can itself be equipped with an abelian group structure and, as such, it has an action on itself given by left translation [example (3) above] as the n-tuple of real numbers $\mathbf{a} = (a_1, a_2, \ldots, a_n)$ in the group acts on the n-tuple $\mathbf{r} = (r_1, r_2, \ldots, r_n)$ by

$$\mathbf{r} = \begin{pmatrix} r_1 \\ r_2 \\ \vdots \\ r_n \end{pmatrix} \rightsquigarrow L_{\mathbf{a}}(\mathbf{r}) := \begin{pmatrix} r_1 \\ r_2 \\ \vdots \\ r_n \end{pmatrix} + \begin{pmatrix} a_1 \\ a_2 \\ \vdots \\ a_n \end{pmatrix},$$

i.e. $r'_i := r_i + a_i$. \hfill (4.10)

We can combine these two actions and get an action of the pair $(\mathbf{a}, A)$ on the Cartesian space $\mathbb{R}^n$ in the form

$$L_{(\mathbf{a}, A)} \mathbf{r} := \mathbf{a} + A\mathbf{r}$$

or, \hfill (4.11)

$$(L(\mathbf{a}, A) \mathbf{r})_i = a_i + \sum_{j=1}^{n} A_{ij} r_j, \text{ for } i = 1, \ldots, n.$$

It might appear at a first glance that this gives an action of the product group $\mathbb{R}^n \times GL(n, \mathbb{R})$ on $\mathbb{R}^n$, but in fact this is not so. To see this, let us perform two consecutive transformations of the type (4.11) with group parameters $(\mathbf{a}_1, A_1)$ and $(\mathbf{a}_2, A_2)$. This gives the (matrix form) equation

$$(L_{(\mathbf{a}_2, A_2)} \circ L_{(\mathbf{a}_1, A_1)})\mathbf{r} = L_{(\mathbf{a}_2, A_2)}(\mathbf{a}_1 + A_1\mathbf{r}) = \mathbf{a}_2 + A_2(\mathbf{a}_1 + A_1\mathbf{r})$$

$$= (\mathbf{a}_2 + A_2\mathbf{a}_1) + A_2A_1\mathbf{r} \tag{4.12}$$

$$= L_{(\mathbf{a}_2 + A_2\mathbf{a}_1, A_2A_1)}(\mathbf{r}) \, .$$

Thus,

$$L_{(\mathbf{a}_2, A_2)} \circ L_{(\mathbf{a}_1, A_1)} = L_{(\mathbf{a}_2 + A_2\mathbf{a}_1, A_2A_1)} \tag{4.13}$$

which cannot be an action of the product group $\mathbb{R}^n \times \text{GL}(n, \mathbb{R})$ since this has the composition law $(\mathbf{a}_2, A_2)(\mathbf{a}_1, A_1) = (\mathbf{a}_2 + \mathbf{a}_1, A_2A_1)$. In fact, Eq. (4.13) represents an action of what is called the *semi-direct product* of the abelian group $\mathbb{R}^n$ and the general linear group $\text{GL}(n, \mathbb{R})$. The composition law for this new group is defined to be

$$(\mathbf{a}_2, A_2)(\mathbf{a}_1, A_1) := (\mathbf{a}_2 + A_2\mathbf{a}_1, A_2A_1) \tag{4.14}$$

so that Eq. (4.13) does indeed define an action on the Cartesian space $\mathbb{R}^n$. Various notations are used to indicate the semi-direct product group structure; I shall write it as $\mathbb{R}^n \, \text{Ⓢ} \, \text{GL}(n, \mathbb{R})$.

From the perspective of applications to physics, a particularly important subgroup of $\mathbb{R}^n \, \text{Ⓢ} \, \text{GL}(n, \mathbb{R})$ is the *Euclidean group* $\mathbb{R}^n \, \text{Ⓢ} \, \text{O}(n, \mathbb{R})$ which is defined by restricting the matrices A to lie in the $\text{O}(n, \mathbb{R})$ subgroup of $\text{GL}(n, \mathbb{R})$. A crucial feature of the orthogonal group $\text{O}(n, \mathbb{R})$ in its action on $\mathbb{R}^n$ as in Eq. (4.9) is that it preserves the 'dot product' between two 'vectors' in the Cartesian space $\mathbb{R}^n$. i.e., if $\mathbf{r}$ is defined as in Eq. (4.9), but with A belonging to $\text{O}(n, \mathbb{R})$, then

$$\sum_{i=1}^{n} r'_{1i}\, r'_{2i} = \sum_{i=1}^{n} r_{1i}\, r_{2i}\,. \tag{4.15}$$

We shall return later to this property of the orthogonal group in a more general context. But to obtain some immediate insight into the role of the Euclidean group in physics let us consider the relation between the coordinates assigned to a vector $\mathbf{v}$ in ordinary 3-space by two observers O and O' whose orthogonal frames of reference are related as shown in the diagram below.

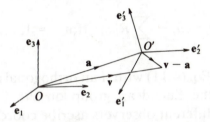

Observer O employs as basis vectors the orthogonal triad $\mathbf{e}_1, \mathbf{e}_2, \mathbf{e}_3$ with $\mathbf{e}_i \cdot \mathbf{e}_j = \delta_{ij}$, while observer O' uses the orthogonal triad $\mathbf{e}'_1, \mathbf{e}'_2, \mathbf{e}'_3$ which also satisfy $\mathbf{e}'_i \cdot \mathbf{e}'_j = \delta_{ij}$ and which are displaced from the origin of the triad used by O by a vector $\mathbf{a}$. Then the coordinates assigned by O and O', respectively to the same abstract vector $\mathbf{v}$ are

For O: $(v_1, v_2, v_3) = (\mathbf{e}_1 \cdot \mathbf{v},\ \mathbf{e}_2 \cdot \mathbf{v},\ \mathbf{e}_3 \cdot \mathbf{v})\,,$ (4.16)

For O': $(v'_1, v'_2, v'_3) = (\mathbf{e}'_1 \cdot (\mathbf{v} - \mathbf{a}),\ \mathbf{e}'_2 \cdot (\mathbf{v} - \mathbf{a}),\ \mathbf{e}'_3 \cdot (\mathbf{v} - \mathbf{a}))\,.$

(4.17)

Now, since $\{\mathbf{e}_1, \mathbf{e}_2, \mathbf{e}_3\}$ is a basis for the 3-space it follows that, if we 'mentally transport' the triad $\mathbf{e}'_1, \mathbf{e}'_2, \mathbf{e}'_3$ back to the origin O, it must be possible to expand each of these three vectors in terms of the triad $\mathbf{e}_1, \mathbf{e}_2, \mathbf{e}_3$. Thus there exists some 3×3 real matrix R_{ij} such that

$$\mathbf{e}'_i = \sum_{j=1}^{3} \mathbf{e}_j R_{ji} \quad \text{for } i = 1, \ldots, 3 \qquad (4.18)$$

and the condition that $\mathbf{e}'_i \cdot \mathbf{e}'_j = \delta_{ij}$ shows at once that R must in fact be an orthogonal matrix, i.e., an element of the group O(3, $\mathbb{R}$). Substituting Eq. (4.18) into Eq. (4.17), we find that $v'_i = \Sigma_{i=1}^{3}(v_i - a_i)R_{ij}$ so that the coordinates assigned to the vector $\mathbf{v}$ by the two observers O and O' are related by

$$v_i = a_i + \sum_{j=1}^{3} R_{ij} v'_j \,; \text{ for } i = 1, \ldots, 3 \,. \qquad (4.19)$$

Hence, from Eq. (4.11) with A an orthogonal matrix, we see that the action of the Euclidean group on the Cartesian space $\mathbb{R}^3$ describes how different observers ascribe coordinates to the same abstract vector. It is not surprising therefore that this group $\mathbb{R}^n \textcircled{s} O(3, \mathbb{R})$ plays an important role in the more mathematical aspects of the structure of Newtonian mechanics and in its quantization.

We have seen that a natural space on which a group can act is itself with the action being left or right (for a right action) multiplication. Another example of this type, and one with far reaching consequences, is when a group is acted upon by one of its subgroups.

Definition.

Let H be a subgroup of a group G. Then a natural right action of H on G can be defined by

$$R_h(g) := gh \,. \qquad (4.20)$$

If H is a proper subgroup of G, this action will not be transitive and the orbits of H in G are called the *left cosets* of H in G. The orbit through a particular element g in G is usually written as gH and the set of all left cosets is denoted G/H.

There is a similar definition of a *right coset* and these are written as Hg with the set of all such cosets being H/G.

Note. (1) The orbit through an element g in G is the same set as the orbit through the point gh for all h in H. Thus, for all h in H, we have $ghH = gH$.

(2) If two group elements g_1 and g_2 lie on the same orbit of H (i.e., if $g_1H = g_2H$) then there must exist some h in H such that $g_2 = g_1h$, i.e., $g_1^{-1}g_2$ belongs to the subgroup H.

(3) Since $gh_1 = gh_2$ implies that $h_1 = h_2$, it follows that each orbit of H in G has the same number of elements which, in the case that H is a finite group, is just the order $|H|$ of H, i.e., the number of elements in H.

This last observation leads at once to a useful and well-known result due originally to Lagrange.

Theorem (Lagrange).

The order $|G|$ of a finite-order group G is an integral multiple of the order $|H|$ of any of its subgroups H.

Proof.

Consider the left cosets of H in G. Each coset/orbit of H in G has $|H|$ elements [by the observation in (3) above] and any two orbits are disjoint with no elements in common.

Furthermore, each element g in G must lie on *some* H-orbit and hence G decomposes into a union of disjoint sets, each of which has $|H|$ elements. Hence, the number of elements in G must equal the number of elements in H multiplied by the number n of H cosets in G, i.e., $|G| = n|H|$.

Corollary.

A group whose order is a prime number is necessarily one of the cyclic groups $\mathbb{Z}_m$.

Proof.

Let g be an element in G and consider the subset $\{e, g, g^2, \ldots, g^{m-1}\}$, where m is the order of g, i.e., the smallest integer such that $g^m = e$. Then this subset is in fact a subgroup of G and is isomorphic to the group $\mathbb{Z}_m$. If we consider the cosets of $\mathbb{Z}_m$ in G, it follows from Lagrange's theorem that $|G| = |\mathbb{Z}_m| n$, where n is the number of cosets; i.e., $|G| = mn$. But $|G|$ is a prime number and has no divisors other than the number 1. Thus either m or n is equal to 1. By choosing $g \neq e$, we can assure that $m \neq 1$, and hence the number of cosets n is one and $m = p$, the prime order of G. Thus $G \cong \mathbb{Z}_p$.

QED.

Note. This proof also shows that, for any finite-order group G, the order of any element in G is necessarily a divisor of the order $|G|$ of G.

The coset space G/H is the set of all orbits of the natural right action of H on G. It is a fact of considerable importance that this space G/H itself carries a natural left action of the group G defined by

$$l_g(g'H) := gg'H \qquad (4.21)$$

so that the action of an element g in G is such as to permute the right H-orbits/cosets. It is clear that this action is transitive since if

g_1H and g_2H are two disjoint H-orbits (and hence two different points in the space G/H) then $l_{(g_1g_2^{-1})}(g_2H) = g_1H$.

What is rather remarkable, and exceptionally useful in practice, is that the *converse* statement is also true, i.e., if X is a space on which a group acts transitively then X is essentially of the form G/H for some subgroup H of G. The precise statement and proof of this result is as follows.

Theorem.

Let X be a set on which a group G acts transitively. Then there exists some subgroup H of G such that there is a bijection $i : G/H \to X$ which preserves the group action in the sense of Eq. (4.4). Thus, for all g in G, we have

$$
\begin{array}{ccc}
G/H & \xrightarrow{\ i\ } & X \\
\downarrow{\scriptstyle l_g} & & \downarrow{\scriptstyle L_g} \\
G/H & \xrightarrow[\ i\ ]{} & X
\end{array}
\qquad i \circ l_g = L_g \circ i, \qquad (4.22)
$$

where L_g denotes the G-action on X and l_g is the special G-action on G/H as defined in Eq. (4.21).

Proof.

Fix some point x_0 in X and let $G_{x_0} := \{g$ in G such that $gx_0 = x_0\}$. Thus G_{x_0} is the set of all elements in G which do not move the point x_0 in X. This rather important subset is in fact a subgroup of G (Exercise) and is variously known as the *stability group/stabilizer group/isotropy group/little group* of G at the point x_0.

We choose G_{x_0} for H and define a map $i : G/H \to X$ by $i(gH) := gx_0$. Then,

(i)   The map is well-defined in this way since if $gH = g'H$ then there exists some h in H such that $g = g'h$ and so $gx_0 = g'hx_0 = g'x_0$ since h belongs to the stability group of the point x_0.

(ii) The map i is one-to-one since if $i(gH) = i(g'H)$ for some g and g' in G then $gx_0 = g'x_0$ by the definition of i. But then $x_0 = g^{-1}g'x_0$ and so $g^{-1}g'$ belongs to $G_{x_0} = H$. Thus there exists some element h in H such that $g' = gh$ and hence it follows that $g'H = gH$.

(iii) The map i is surjective (i.e., onto) since every x in X is of the form $x = gx_0$ for some g in G because the G-action on X is transitive.

So far, we have shown that the set X and the coset space G/G_{x_0} can be placed in a bijective, one-to-one correspondence. Now we must show that the map i preserves the group action in the sense of Eq. (4.22). But, for any g' in G, we have

$$L_g \circ i(g'H) = L_g(i(g'H)) = L_g(g'x_0)$$
$$= gg'x_0 = i(gg'H) = i \circ l_g(g'H)$$

which, since it is true for any g' in G, implies $L_g \circ i = i \circ l_g$.

$$\text{QED.}$$

Note. Any G-action on any set X decomposes X into a disjoint union of G-orbits, on each one of which the action is transitive. From the above, it follows that each orbit is effectively of the form G/G_x for some arbitrarily chosen point x in the orbit. In this sense, the most general G-action is classifiable in terms of the subgroups of G – a very powerful and useful result.

Example.

Some of the most interesting applications of this theorem arise when G is a Lie group and the space X on which it acts transitively is a "differentiable manifold". For example, consider the (intransitive) action of $SO(3, \mathbb{R})$ on $\mathbb{R}^3$ obtained from Eq. (4.9) as

$$L_A \begin{pmatrix} r_1 \\ r_2 \\ r_3 \end{pmatrix} := \begin{pmatrix} A_{11} & A_{12} & A_{13} \\ A_{21} & A_{22} & A_{23} \\ A_{31} & A_{32} & A_{33} \end{pmatrix} \begin{pmatrix} r_1 \\ r_2 \\ r_3 \end{pmatrix}. \tag{4.23}$$

The fact that A belongs to $SO(3, \mathbb{R})$, i.e., $AA^t = \mathbf{1}$, is equivalent to the statement that the action in Eq. (4.23) leaves fixed the dot product between two vectors in $\mathbb{R}^3$. In particular, the length of a vector is unchanged by this action and hence if we define the unit 2-sphere as

$$S^2 := \{\mathbf{r} \text{ in } \mathbb{R}^3 \text{ such that } \mathbf{r} \cdot \mathbf{r} = 1\} \tag{4.24}$$

then this space is mapped into itself under the $SO(3, \mathbb{R})$ action.

It is obvious "geometrically" (Exercise: prove it analytically) that any unit vector can be mapped into any other unit vector by a $SO(3, \mathbb{R})$ action. Thus the induced action on S^2 is transitive, and the previous theorem is applicable.

It follows that S^2 can be put in bijective correspondence with the coset space $SO(3, \mathbb{R})/H$ where H is the stability group of any conveniently chosen "fiducial" vector $\mathbf{r}_0$ in $\mathbb{R}^3$. A particularly easy choice to handle is the unit vector $\begin{pmatrix} 0 \\ 0 \\ 1 \end{pmatrix}$ which points along the "z-axis". But we note

$$\left(\begin{array}{c|c} X & \begin{matrix} 0 \\ 0 \end{matrix} \\ \hline 0 \quad 0 & 1 \end{array} \right) \begin{pmatrix} 0 \\ 0 \\ 1 \end{pmatrix} = \begin{pmatrix} 0 \\ 0 \\ 1 \end{pmatrix} \tag{4.25}$$

for any choice of 2×2 matrix X. Also, the 3×3 matrix on the left hand side of Eq. (4.25) will belong to SO(3, $\mathbb{R}$) if and only if X is a 2×2 orthogonal, unit determinant matrix. Furthermore, the only matrices in SO(3, $\mathbb{R}$) that satisfy Eq. (4.25) are of this special form.

It follows from all this that the isotropy group of the fiducial vector $\mathbf{r}_0$ is the subgroup SO(2, $\mathbb{R}$) (i.e., the rotations about the z-axis) and hence, from the theorem, that there is a bijective correspondence between S^2 and the coset space SO(3, $\mathbb{R}$)/SO(2, $\mathbb{R}$).

Note. Considered as manifolds, the Lie groups SO(3, $\mathbb{R}$) and SO(2, $\mathbb{R}$) have dimension 3 and 1, respectively, while $\dim(S^2) = 2$. This is a special case of a very important general result: If H is a Lie subgroup of a Lie group G, then the coset space G/H can itself be made into a differentiable manifold, and in such a way that $\dim(G/H) = \dim(G) - \dim(H)$.

1.5. NORMAL SUBGROUPS

We have seen that the space G/H of left cosets/orbits of the right H-action on G has $|G|/|H|$ elements and that G acts on it on the left as a group of transformations. A natural question that arises is whether G/H is itself a group, obtained as it were by 'dividing' G by H. One obvious way of trying to impose a group structure on G/H is to define a composition of the cosets $g_1 H$ and $g_2 H$ as

$$(g_1 H)(g_2 H) := g_1 g_2 H \qquad ? \qquad (5.1)$$

However, in order for this to make sense, the right hand side should be independent of the particular way in which the cosets on the left hand side are represented. For example, we know that $g_1 H$ and $g_1 h H$ are the same coset (and hence represent the same point in G/H) for any h in H. But if we use $g_1 h H$ in the left hand side of Eq. (5.1), right hand side becomes $g_1 h g_2 H$ and in general this coset will *not* be the same as $g_1 g_2 H$. However, for special choices of the subgroup H, everything will work correctly, and these are the so-called "normal" subgroups.

Definition.

A subgroup H of a group G is said to be *normal* (or *invariant*) if for any pair h in H and g in G, there exists a h' in H such that

$$ghg^{-1} = h' . \tag{5.2}$$

In this case, Eq. (5.1) *is* well-defined because, for example, we now have that $g_1 h g_2 = g_1 g_2 h'$ for some h' in H and hence $g_1 h g_2 H = g_1 g_2 H$ as was required for consistency. [Exercise: Show that (5.1) defines a group law.]

Note. (1) Condition (5.2) is equivalent to saying that $gH = Hg$ for all g in G; i.e., the left and right cosets of g are equal.

(2) Any subgroup of an abelian group G is necessarily normal since $hg = gh$ and hence one can choose $h' = h$ in Eq. (5.2). Hence, if G is abelian, the coset space is a group (itself abelian) for any subgroup H of G.

(3) If H is normal then Lagrange's theorem shows that the order $|G/H|$ of the group G/H is $|G|/|H|$.

Examples.

(1) Consider the $\mathbb{Z}_2$ subgroup, $\{e, c\}$ of the order-4 group V_4 ($\cong \mathbb{Z}_2 \times \mathbb{Z}_2$) whose group table was given in Eq. (2.6). In accord with Lagrange's theorem, there are just two cosets, $e\mathbb{Z}_2$ and $a\mathbb{Z}_2$, corresponding to the orbits $O_e = \{e, c\}$ and $O_a = \{a, b\}$. Since V_4 is abelian, the subgroup $\mathbb{Z}_2$ is normal and $V_4/\mathbb{Z}_2$ is itself a group with $(a\mathbb{Z}_2)(a\mathbb{Z}_2) = e\mathbb{Z}_2$; i.e., it is isomorphic to $\mathbb{Z}_2$, as one would expect for an order-2 group.

(2) Let $\mathbb{Z}_2 := \left\{ \begin{pmatrix} 1 & 0 \\ 0 & 1 \end{pmatrix}, \begin{pmatrix} -1 & 0 \\ 0 & -1 \end{pmatrix} \right\}$ be considered as a subgroup of SU(2) – the Lie group of all 2×2 unitary matrices with determinant equal to one. Then this particular $\mathbb{Z}_2$ is a normal subgroup since, for all matrices A in SU(2), we have

$$A \begin{pmatrix} -1 & 0 \\ 0 & -1 \end{pmatrix} = \begin{pmatrix} -1 & 0 \\ 0 & -1 \end{pmatrix} A$$

and of course it is trivial that the unit matrix commutes with all matrices A in $SU(2)$.

This particular example is intriguing for a number of reasons. From a topological perspective, the 'factoring out' by $\mathbb{Z}_2$ is equivalent to identifying antipodal points on the 3-sphere that is the group space of $SU(2)$. The resulting manifold is rather difficult to picture: it is known technically as the three-dimensional real projective space $\mathbb{RP}^3$.

Since $\mathbb{Z}_2$ is a normal subgroup of $SU(2)$, it follows that the coset space $SU(2)/\mathbb{Z}_2$ must itself be a group, and in fact it can be shown that it is isomorphic to the 3-dimensional special orthogonal group; i.e.,

$$SU(2)/\mathbb{Z}_2 \cong SO(3, \mathbb{R}) \qquad (5.3)$$

which shows, in particular, that the topological structure of this 'rotation group' $SO(3, \mathbb{R})$ is that of the projective space $\mathbb{RP}^3$. Both the topological and the group theoretic connections between the groups $SU(2)$ and $SO(3, \mathbb{R})$ are of considerable importance in understanding the phenomenon of 'intrinsic spin' in quantum theory (as in the spin-$\frac{1}{2}$ of the electron).

(3) Let $n\mathbb{Z} := \{nm,$ where m is in $\mathbb{Z}\}$ be a subgroup of the abelian group of integers $\mathbb{Z}$. Thus, for example,

$$2\mathbb{Z} = \{\ldots, -6, -4, -2, 0, 2, 4, 6, \ldots\}$$
$$3\mathbb{Z} = \{\ldots, -12, -9, -3, 0, 3, 6, 9, \ldots\}$$

and so on. Then $\mathbb{Z}/n\mathbb{Z}$ must itself be a group since $n\mathbb{Z}$ is a subgroup of the abelian group $\mathbb{Z}$. It is easy to see that if $\mathbb{Z}_n := \{e, a, a^2, \ldots, a^{n-1}\}$ with $a^n = e$, then $\mathbb{Z}_n$ is isomorphic to

$\mathbb{Z}/n\mathbb{Z}$ via a map that takes a^m into the coset $m(n\mathbb{Z})$ for each m such that $0 \le m \le n - 1$. Note that two integers lie in the same coset if, and only if, they differ by a multiple of n. This is often expressed by saying that they are "equal modulo n". It is the existence of the isomorphism $\mathbb{Z}/n\mathbb{Z} \cong \mathbb{Z}_n$ that is mainly responsible for the notation "$\mathbb{Z}_n$".

We come now to one of the most important applications of the idea of a normal subgroup. This concerns the situation when we have a homomorphism [Eq. (4.1)] between a pair of groups G_1 and G_2 and the extent to which this can be regarded as a "faithful" representation of G_1 in G_2. The key concepts are as follows

Definitions.

Let $\mu : G_1 \to G_2$ be a homomorphism between the two groups G_1 and G_2. Then,

(a) The *image* of μ (written as Im μ) is the set of all elements in G_2 which are mapped into by μ, i.e.,

$$\text{Im } \mu := \left\{ \begin{array}{l} g_2 \text{ in } G_2 \text{ such that there exists } g_1 \text{ in } G_1 \\ \text{with } \mu(g_1) = g_2 \end{array} \right\}.$$

(b) The *kernel* of μ (written as Ker μ) is the set of all elements in G_1 that are mapped by μ into the unit element e_2 in G_2. i.e.,

$$\text{Ker } \mu := \{ g \text{ in } G_1 \text{ such that } \mu(g) = e_2 \}$$

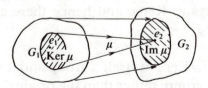

Note. (1) The image of μ is a subgroup of G_2. (Exercise)

(2) The kernel of μ is the subgroup (Exercise) of G_1 that

'fails to be represented' by μ. Indeed, let g and g' be any pair of elements in G_1 for which $\mu(g) = \mu(g')$; i.e., the homomorphism cannot distinguish between them and they are 'represented' by the same element in G_2. Then

$$e_2 = [\mu(g)]^{-1} \mu(g') = \mu(g^{-1})\mu(g') = \mu(g^{-1}g')$$

and so $g^{-1}g'$ belongs to Ker μ. Thus there exists some k in Ker μ such that $g' = gk$ and, in this sense, g and g' "differ" by an element of Ker μ.

Conversely, $\mu(gk) = \mu(g)\,\mu(k) = \mu(g)$ if k is in Ker μ. Thus the extent to which μ fails to be one-to-one is captured precisely by the kernel.

This observation that μ maps all members of the coset $g(\text{Ker }\mu)$ to the same element in G_2 makes the following theorem particularly interesting.

Theorem.

The kernel of μ, Ker μ, is a normal subgroup of G_1.

Proof.

Let k and g belong to Ker μ and G_1, respectively. Then,

$$\mu(gkg^{-1}) = \mu(g)\,\mu(k)\,\mu(g^{-1}) = \mu(g)\,e_2\mu(g^{-1})$$
$$= \mu(g)\,\mu(g^{-1}) = \mu(gg^{-1}) = \mu(e_1) = e_2\,.$$

Thus gkg^{-1} belongs to Ker μ and hence there exists some k' in Kerμ such that $gkg^{-1} = k'$.

QED.

It follows that $G_1/\text{Ker }\mu$ is itself a group and it seems intuitively clear that it is *this* group, rather than the original group G_1, that is "faithfully represented" in the group G_2. The precise statement of this result is the following:

Theorem.

The homomorphism μ of G_1 into G_2 induces an *isomorphism* between $G_1/\text{Ker}\,\mu$ and the subgroup $\text{Im}\,\mu$ of G_2.

Proof.

Define $i: G_1/\text{Ker}\,\mu \rightarrow \text{Im}\,\mu$ by $i(gK) := \mu(g)$. (For typographical ease, I am writing K for $\text{Ker}\,\mu$.) The observation that μ maps all members of the coset $g(\text{Ker}\,\mu)$ into the same element in G_2 shows that this map i is well-defined.

Suppose then that $i(gK) = i(g'K)$ for some g and g' in G_1. Then $\mu(g) = \mu(g')$ and, as shown in (2) above, this implies that there exists some k in K such that $g' = gk$. But then $gK = g'K$, and hence μ is one-to-one. It is trivially a map onto $\text{Im}\,\mu$, and we conclude that the map $i: G_1/\text{Ker}\,\mu \rightarrow \text{Im}\,\mu$ is a bijection.

To show that it is an isomorphism, we must prove that it preserves the group law on the quotient group $G_1/\text{Ker}\,\mu$. But,

$$i(gK\,g'K) = i(gg'K) = \mu(gg') = \mu(g)\,\mu(g')$$
$$= i(gK)\,i(g'K).$$

QED.

This result is very useful in a variety of applications. One frequently occurring situation is when the map μ is known to be a *surjective* homomorphism onto G_2. It then follows from the theorem that $G_1/\text{Ker}\,\mu$ is isomorphic to G_2 itself; this is quite a common way of showing that two groups are isomorphic. For example:

Corollary.

There is a $\mathbb{Z}_2$ subgroup of the Lie group $SL(2, \mathbb{C})$ such that the group of Möbius transformations [see Eq. (3.1)] is isomorphic to $SL(2, \mathbb{C})/\mathbb{Z}_2$.

Proof.

Define a map μ from $SL(2, \mathbb{C})$ to the group of Möbius transformations by associating with the matrix $\begin{pmatrix} a & b \\ c & d \end{pmatrix}$, the Möbius transformation

$$z \rightsquigarrow \frac{az + b}{cz + d}.$$

This is a homomorphism (Exercise) and it is clearly surjective. The kernel of μ is the set of all matrices $\begin{pmatrix} a & b \\ c & d \end{pmatrix}$ such that

$$z = \frac{az + b}{cz + d} \qquad \text{for all } z \text{ in } \mathbb{C}.$$

Thus $cz^2 + dz = az + b$ and this can only be true for all z if $c = b = 0$ and $d = a$. But the matrix belongs to the special linear group and hence has determinant equal to one, which implies that $ad = 1$. Hence the only two possibilities are $a = d = +1$ and $a = d = -1$. Thus Ker μ contains just the two matrices

$$\text{Ker } \mu = \left\{ \begin{pmatrix} 1 & 0 \\ 0 & 1 \end{pmatrix}, \begin{pmatrix} -1 & 0 \\ 0 & -1 \end{pmatrix} \right\}$$

which is a $\mathbb{Z}_2$ group. Since the homomorphism μ is surjective, it follows from the theorem that $SL(2, \mathbb{C})/\mathbb{Z}_2$ is isomorphic with the Möbius group.

QED.

2. VECTOR SPACES

2.1. BASIC DEFINITIONS

The two fundamental algebraic properties of ordinary three-dimensional vectors are:

(i) Any pair of vectors **v** and **w** can be added together to give a third vector **v** + **w**;

(ii) A vector **v** can be multiplied by a real number r to get a new vector r**v** and this action is linear in the sense that $r(\mathbf{v} + \mathbf{w}) = r\mathbf{v} + r\mathbf{w}$ for any pair of vectors **v** and **w**.

It is clear that the set of all such 3-dimensional vectors forms an abelian group under the "+" operation (with the unit element being the null vector **0**) and that this group structure is 'augmented' in some way by the additional property of being able to multiply vectors by a real number.

These algebraic properties of ordinary vectors form one of the motivations underlying the definition of a general vector space — the topic with which the second part of this course is concerned. It is important to introduce the idea at once of a "complex" vector space in which complex, rather than real, numbers are used to multiply vectors. Vector spaces of this type play a fundamental role in quantum theory (the state of a quantum system is represented by a vector in a complex space) and in the theory of group representations.

Definition.

A *complex vector space V* is an abelian group plus an additional operation known as *scalar multiplication* which associates with each complex number μ and vector **v**, a new vector written μ**v** and satisfying the conditions:

(a) $\mu(\mathbf{v}_1 + \mathbf{v}_2) = \mu\mathbf{v}_1 + \mu\mathbf{v}_2$,　　　　　　　　　　　　　　(1.1)

(b) $(\mu_1 + \mu_2)\mathbf{v} = \mu_1\mathbf{v} + \mu_2\mathbf{v}$,　　　　　　　　　　　　　　(1.2)

(c) 　$\mu_1(\mu_2\mathbf{v}) = (\mu_1\mu_2)\mathbf{v}$,　　　　　　　　　　　　　　　(1.3)

(d) 　　$1\,\mathbf{v} = \mathbf{v}$,　　　　　　　　　　　　　　　　　(1.4)

(e) 　　$0\,\mathbf{v} = \mathbf{0}$,　　　　　　　　　　　　　　　　　(1.5)

for all complex numbers μ, μ_1 and μ_2, and all vectors **v**, $\mathbf{v}_1$ and $\mathbf{v}_2$.

Note. (1) The "0" on the left hand side of Eq. (1.5) is the number zero, whereas the "**0**" on the right hand side refers to the null vector in V, i.e., the unit element of V regarded as an abelian group.

(2) It is conventional to write $(-1)\mathbf{v}$ as $-\mathbf{v}$ and $\mathbf{v} + (-\mathbf{w})$ as $\mathbf{v} - \mathbf{w}$.

(3) There is a completely analogous definition of a *real vector space* in which conditions (1.1–5) are unchanged except that real, rather than complex, numbers are used.

In Sec. 1.5, we introduced the idea of a homomorphism between two groups as a map that preserved the group law in the sense of Eq. (1.5.4). This is in fact a special example of the far more general concept of a "morphism" which is defined roughly as a map between any two spaces equipped with a 'mathematical structure' which preserves that structure.

In the case of vector spaces, a morphism must not only preserve the additive law (i.e., it must be a homomorphism between the underlying abelian groups) but it must also respect the scalar multiplication. Such a map is said to be 'linear' and is defined formally as follows

Definitions.

(a) A *linear map* between two vector spaces V_1 and V_2 is a map $L: V_1 \to V_2$ which preserves the vector space structure in the sense that:

$$L(\mu_1 \mathbf{v}_1 + \mu_2 \mathbf{v}_2) = \mu_1 L(\mathbf{v}_1) + \mu_2 L(\mathbf{v}_2) \qquad (1.6)$$

for all complex numbers μ_1, μ_2 and vectors $\mathbf{v}_1$, $\mathbf{v}_2$.

(b) An *isomorphism* is a linear map that is also one-to-one and sur-jective. (This implies that the inverse map $L^{-1}: V_2 \to V_1$ is also linear.) If there is an isomorphism between two vector spaces V_1 and V_2 then V_1 and V_2 are said to be *isomorphic*.

(c) An isomorphism of a vector space V with itself is called an *automorphism* of V. Note that the set of all automorphisms of a vector space is a group, denoted $\mathrm{Aut}(V)$.

Examples.

(1) The abelian group $\mathbb{C}^n$ may be given the structure of a complex vector space by defining the scalar multiplication as

$$\mu(a_1, a_2, \ldots, a_n) := (\mu a_1, \mu a_2, \ldots, \mu a_n). \qquad (1.7)$$

Similarly, the abelian group $\mathbb{R}^n$ can be made into a real vector space. [Exercise. Show that (1.7) does define a vector space structure on $\mathbb{C}^n$.]

(2) The abelian group $M(n, \mathbb{C})$ of all $n \times n$ complex matrices can be given the structure of a complex vector space by defining the scalar multiplication as

$$(\mu A)_{ij} := \mu A_{ij} \qquad \text{for all } i, j = l, \ldots, n, \qquad (1.8)$$

where μ is any complex number and A belongs to $M(n, \mathbb{C})$.

Exercise. Show that the vector spaces $\mathbb{C}^{n^2}$ and $M(n, \mathbb{C})$ are isomorphic.

(3) Let $\mathbb{C}^\infty$ denotes the set of all infinite sequences $(a_1, a_2, \ldots)$ of complex numbers. This can be made into a complex vector space by a procedure strictly analogous to the one employed in the case of $\mathbb{C}^n$.

(4) Let X be any set and let V be a complex vector space. Then the set Map(X, V) of all maps from X to V can be given the structure of a complex vector space by defining

$$(f_1 + f_2)(x) := f_1(x) + f_2(x) \qquad (1.9)$$

$$(\mu f)(x) := \mu(f(x)) \qquad (1.10)$$

for all f_1 and f_2 in Map(X, V). Note that the vector space structure of V appears explicitly in the right hand side of Eqs. (1.9–10). The simplest example is when $V = \mathbb{C}$ and, indeed, the spaces Map$(X, \mathbb{C})$ appear frequently in many areas of mathematics.

A linear map is the vector space analogue of the group theoretic concept of a homomorphism. Many other constructs in group theory have equivalents in vector space theory which are defined by using the original construct as far as the underlying abelian group is concerned and then augmenting it to make it consistent with the additional action of scalar multiplication.

In particular, the vector space analogue of the idea of a subgroup is contained in the concept of a 'subspace'.

Definition.

A subset W of a vector space V is said to be a *linear subspace* of V if:

(a) W is a subgroup of V with respect to the abelian group structure, i.e., under the '+' operation.

(b) W is also 'closed' under scalar multiplication, i.e., if μ is in $\mathbb{C}$ and **w** belongs to W, then μ**w** also lies in W.

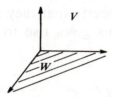

Note. If W is a linear subspace of V then it is in particular a normal subgroup of the underlying abelian group and hence V/W is also an abelian group. This quotient space can be given a vector space structure by defining the scalar multiplication as:

$$\mu(\mathbf{v} + W) := \mu\mathbf{v} + W, \qquad (1.11)$$

where $\mathbf{v} + W$ denotes the coset of **v** with respect to the abelian subgroup W. (Exercise. Show that this *does* define a vector space.)

Examples.

(1) In $\mathbb{C}^n$, the set of all n-tuples of complex numbers of the form $(a_1, a_2, \ldots, a_m, 0, 0, \ldots, 0)$ with $m < n$, is a linear subspace that is clearly isomorphic to the vector space $\mathbb{C}^m$.
Similarly, for any finite n, the vector space $\mathbb{C}^n$ can be viewed as a linear subspace of the vector space $\mathbb{C}^\infty$ of all infinite sequences of complex numbers.

(2) A rather interesting (and important) linear subspace of $\mathbb{C}^\infty$ is the set l_2 which is defined to be the set of all sequences $(a_1, a_2, \ldots)$ such that

$$\sum_{i=1}^{\infty} |a_i|^2 < \infty . \qquad (1.12)$$

(3) We can define $C^k(\mathbb{R}^n, \mathbb{C})$ to be the subset of all functions from $\mathbb{R}^n$ to $\mathbb{C}$ with the property that they are continuously differentiable k times. This gives rise to the string of subspace embeddings

$$C^\infty \subset C^k \subset C^{k-1} \subset \ldots \subset C^0 \subset \text{Map}(\mathbb{R}^n, \mathbb{C}) .$$

Another (very important!) subspace of $\text{Map}(\mathbb{R}^n, \mathbb{C})$ is the space $\mathscr{L}^2(\mathbb{R}^n)$ which is defined to be the set of all infinitely differentiable functions that are also square integrable

$$\int_{-\infty}^{\infty} \ldots \int_{-\infty}^{\infty} |f(x)|^2 \, dx^1 \, dx^2 \ldots dx^n < \infty . \qquad (1.13)$$

This is essentially the space of quantum states for a non-relativistic particle moving in n spatial dimensions.

In elementary 3-dimensional vector calculus, it is common to employ basis vectors $\{\mathbf{i}, \mathbf{j}, \text{ and } \mathbf{k}\}$ in terms of which every vector $\mathbf{v}$ can be expanded as $\mathbf{v} = v_x \mathbf{i} + v_y \mathbf{j} + v_z \mathbf{k}$ with coefficients (v_x, v_y, v_z). The extension of this idea to a general (and complex) vector space is of considerable importance.

Definitions.

(a) Let S be a subset (possibly infinite) of the complex vector space V. Then the set of all *finite* linear combinations $\mu_1 v^1 + \mu_2 v^2 + \ldots + \mu_j v^j$ of any set of vectors $\{v^1, v^2, \ldots, v^j\}$ in S (for any finite value of j) is called the *span* $[S]$ of S.

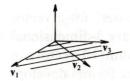

v_1, v_2, and v_3 are coplanar and hence span the plane in which they lie.

(b) A finite set of vectors $\{v^1, v^2, \ldots, v^k\}$ in V is *linearly dependent* if there exists some set of complex numbers $\mu_1, \mu_2, \ldots, \mu_k$ (not all zero) such that

$$\sum_{i=1}^{k} \mu_i v^i = 0. \tag{1.14}$$

If there is no such set of complex numbers (i.e., if the only way of satisfying Eq. (1.14) is for all the μ_i to vanish) then the set of vectors is called *linearly independent*. For example, in the figure above, the sets $\{v^1, v^2\}$, $\{v^1, v^3\}$, and $\{v^2, v^3\}$ are sets of linearly independent vectors whereas the set $\{v^1, v^2, v^3\}$ is a linearly dependent set.

(c) An infinite set of vectors is *linearly independent* if every finite subset of vectors is linearly independent in the sense above.

(d) A vector space is *n-dimensional* (where $n < \infty$) if it contains a subset of n linearly independent vectors but contains no subset of $n + 1$ such vectors.

If it contains n linearly independent vectors for each positive integer n then the vector space is said to have *infinite dimension*.

Examples.

(1) For every $n < \infty$, the vector space $\mathbb{C}^n$ has finite dimension equal to n.
(2) The vector spaces $\mathbb{C}^\infty$, l_2, $C^k(\mathbb{R}^n, \mathbb{C})$ and $\mathscr{L}^2(\mathbb{R}^n)$ are all examples of spaces whose dimension is infinite.

We now come to the important idea of a "basis" for a vector space. Guided by the analogy of **i**, **j**, **k** in ordinary 3-dimensional vector calculus, we expect this to be a set of $\dim(V)$ vectors in terms of which all vectors can be expanded. The formal development of this idea is as follows. (Throughout the rest of this section, V will always be finite-dimensional.)

Definitions.

A subset $S = \{\mathbf{e}^1, \mathbf{e}^2, \ldots, \mathbf{e}^m\}$ of linearly independent vectors of a vector space V is a *basis* for V if $[S] = V$, i.e., any vector **v** in V can be expanded as

$$\mathbf{v} = \sum_{i=1}^{m} v_i \mathbf{e}^i, \quad \text{for some complex numbers } v_1, \ldots, v_m. \quad (1.15)$$

The complex numbers $v_1, v_2, \ldots, v_m$ are called the *expansion coefficients* of the vector **v** with respect to the given basis.

Note. The expansion coefficients are unique. For suppose that

$$\sum_{i=1}^{m} v_i \mathbf{e}^i = \sum_{i=1}^{m} v_i' \mathbf{e}^i = \mathbf{v}.$$

Then,

$$\sum_{i=1}^{m} (v_i - v_i') \mathbf{e}^i = 0$$

which, since $S = \{e^1, e^2, \ldots, e^m\}$ is a linearly independent subset of vectors, implies that $v_i = v_i'$ for all $i = 1, \ldots, m$.

Theorem.

A vector space V is n-dimensional (where $n < \infty$) if and only if it has a basis of n vectors.

Proof.

(a) Suppose that $\{e^1, e^2, \ldots, e^n\}$ is a basis set and let $\{v^1, v^2, \ldots, v^{n+1}\}$ be any set of $n + 1$ vectors. Then there exist complex numbers v_i^j such that

$$v^j = \sum_{i=1}^{n} e^i v_i^j \quad \text{for } j = 1, \ldots, n + 1 .$$

Let $\mu_1, \mu_2, \ldots, \mu_{n+1}$ be any not all-zero solution to the set of n linear equations

$$\sum_{j=1}^{n+1} v_i^j \mu_j = 0 , \quad i = 1, \ldots, n .$$

Then,

$$\sum_{j=1}^{n+1} \mu_j v^j = \sum_{j=1}^{n+1} \mu_j \left(\sum_{i=1}^{n} e^i v_i^j \right) = \sum_{i=1}^{n} e^i \left(\sum_{j=1}^{n+1} v_i^j \mu_j \right) = 0 .$$

Hence $\{v^1, v^2, \ldots, v^{n+1}\}$ is a linearly dependent set and so V is n-dimensional.

(b) Conversely, if V is n-dimensional, there exists a subset $\{e^1, e^2, \ldots, e^n\}$ of linearly independent vectors. Then for any v in V, the subset $\{v, e^1, e^2, \ldots, e^n\}$ must be linearly dependent and hence there exist complex numbers $\mu_1, \mu_2, \ldots, \mu_n, \mu_{n+1}$, not all zero, such that

$$\mu_1 \mathbf{e}^1 + \mu_2 \mathbf{e}^2 + \ldots + \mu_n \mathbf{e}^n + \mu_{n+1} \mathbf{v} = 0 . \qquad (1.16)$$

But $\{\mathbf{e}^1, \mathbf{e}^2, \ldots, \mathbf{e}^n\}$ are linearly independent and hence $\mu_{n+1} \neq 0$. Thus Eq. (1.16) can be solved for $\mathbf{v}$ as $\mathbf{v} = -\mu_{n+1}^{-1} \Sigma_{i=1}^{n} \mu_i \mathbf{e}^i$, so that $\{\mathbf{e}^1, \mathbf{e}^2, \ldots, \mathbf{e}^n\}$ is a basis set.

QED.

Corollary.

Any n-dimensional $(n < \infty)$ complex vector space is isomorphic to the vector space $\mathbb{C}^n$.

Proof.

Let $\{\mathbf{e}^1, \mathbf{e}^2, \ldots, \mathbf{e}^n\}$ be a basis set for the n-dimensional vector space V and let $\mathbf{v} = \Sigma v_i \mathbf{e}^i$ be the unique expansion of $\mathbf{v}$ in V with respect to this basis. Then we define a map $i : V \to \mathbb{C}^n$ by $i(\mathbf{v}) := (v_1, v_2, \ldots, v_n)$ and it is easy to see that this is indeed an isomorphism. (Exercise. Prove this.)

Examples.

(1) The set $\{(1, \sqrt{3}), (i, 6)\}$ is a basis set for $\mathbb{C}^2$.
(2) For any $n < \infty$, a basis set for $\mathbb{C}^n$ is the n vectors:

$$\{(1, 0, \ldots, 0), (0, 1, \ldots, 0), \ldots, (0, 0, \ldots, 1)\}$$

since any vector $\mathbf{a} = (a_1, a_2, \ldots, a_n)$ can be expanded in terms of these vectors as

$$\mathbf{a} = a_1(1, 0, \ldots, 0) + a_2(0, 1, \ldots, 0) \\ + \ldots + a_n(0, 0, \ldots, 1).$$

2.2. NORMED VECTOR SPACES

Now we wish to see to what extent the results of the previous section can be extended to the case where V is an infinite-dimensional vector space. For example, we might like to consider situations where it is possible to expand any vector $\mathbf{v}$ in V in the form

$$\mathbf{v} = \sum_{i=1}^{\infty} v_i \mathbf{e}^i , \qquad (2.1)$$

where $\{\mathbf{e}^1, \mathbf{e}^2, \ldots\}$ is a "countably infinite" basis set for V. But before we can even start to try and prove infinite-dimensional versions of the finite-dimensional expansion theorems, it is first necessary to decide how an expression such as Eq. (2.1) should be interpreted. What does it mean to write an infinite sum of vectors?

We recall that in the case of complex numbers, $S = \Sigma_{i=1}^{\infty} a_i$ means that S is the limit of the partial sums

$$S := \lim_{N \to \infty} S^N , \quad \text{where } S^N := \sum_{i=1}^{N} a_i \qquad (2.2)$$

and that to say that S is the limit of the sequence $S^1, S^2, \ldots$ of complex numbers means:

"For any $\varepsilon > 0$, there exists $N_0(\varepsilon)$ such that $N > N_0$
implies that $|S - S^N| < \varepsilon$." $\qquad (2.3)$

Motivated by this observation, we will interpret an expression such as Eq. (2.1) to mean that $\mathbf{v} := \lim_{N \to \infty} \Sigma_{i=1}^{N} v_i \mathbf{e}^i$. Which reduces the problem to deciding what it means in a vector space to say that a sequence of vectors $\mathbf{v}^1, \mathbf{v}^2, \ldots$ converges to some limit vector $\mathbf{v}$.

If V is finite-dimensional, we could choose a basis set $\{\mathbf{e}^1, \mathbf{e}^2, \ldots, \mathbf{e}^n\}$ [where $n = \dim(V)$] and say that the sequence of vectors

$\{\mathbf{v}^N\}_{N=1}^{\infty}$ converges to $\mathbf{v}$ if, when we expand $\mathbf{v}^N$ and $\mathbf{v}$ as

$$\mathbf{v}^N = \sum_{i=1}^{n} \mathbf{e}^i v_i^N \qquad \text{and} \qquad \mathbf{v} = \sum_{i=1}^{n} \mathbf{e}^i v_i,$$

we find that, for each $i = 1, \ldots, n$, the sequence of expansion coefficients $v_i^1, v_i^2, \ldots$ converges to the expansion coefficient v_i, in the usual sense of complex numbers.

However, for an infinite-dimensional space, such a procedure is not available (since the problem in the first place was to define infinite sums of vectors) and hence we must proceed in a different way. The key idea is to try and find a direct analogue of the expression (2.3) by finding some suitable analogue in a vector space of the modulus of a complex number, whose "smallness' can measure the "smallness" of a vector. This is provided by the concept of a "norm" of a vector.

Definitions.

(a) A *norm* on a complex vector space is a map $\| \ \ \|$ from V into the finite real numbers satisfying the following conditions:

    (i) $\|\mathbf{v} + \mathbf{w}\| \le \|\mathbf{v}\| + \|\mathbf{w}\|$ for all $\mathbf{v}, \mathbf{w}$ in V.   (2.4)
        (This is the analogue of the "triangle inequality" satisfied by the modulus of complex numbers.)

    (ii) $\|\mu\mathbf{v}\| = |\mu| \ \|\mathbf{v}\|$ for all complex numbers μ and vectors $\mathbf{v}$ in V.   (2.5)

    (iii) $\|\mathbf{v}\| \ge 0$ with $\|\mathbf{v}\| = 0$ only if $\mathbf{v} = 0$.   (2.6)

[Note that, as with (i), the conditions (ii) and (iii) are analogues of properties of the modulus of complex numbers.]

(b) A sequence of vectors $\mathbf{v}^1, \mathbf{v}^2, \ldots$ in a normed vector space V is said to *converge strongly* (or *in the norm*) to a vector $\mathbf{v}$ in V if the sequence of complex numbers $\|\mathbf{v}^N - \mathbf{v}\|$ converges to 0 in the usual way, i.e., for any $\varepsilon > 0$, there exists $N_0(\varepsilon)$ such that $N > N_0$ implies that $\| \mathbf{v}^N - \mathbf{v} \| < \varepsilon.$ (2.7)

We express this by writing "$\mathbf{v} = \lim_{N \to \infty} \mathbf{v}^N$" or "$\mathbf{v}^N \to \mathbf{v}$".

Examples.

(1) On the vector space $\mathbb{C}^n$, we can define a norm on vectors $\mathbf{a} = (a_1, a_2, \ldots, a_n)$ as:

$$\| \mathbf{a} \|^2 := \sum_{i=1}^{n} |a_i|^2 .$$ (2.8)

More generally, on any finite dimensional vector space V, choose a basis set $\{\mathbf{e}^1, \mathbf{e}^2, \ldots, \mathbf{e}^n\}$ and define

$$\| \mathbf{v} \|^2 := \sum_{i=1}^{n} |v_i|^2 , \quad \text{where } \mathbf{v} = \sum_{i=1}^{n} v_i \mathbf{e}^i .$$ (2.9)

It is easy to show that, with this definition of a norm, strong convergence of a sequence of vectors is the same as the convergence of the expansion coefficients mentioned above.

(2) A norm can be defined on the space l_2 of infinite sequences with summable modulus-squared coefficients, by

$$\| \mathbf{a} \|^2 := \sum_{i=1}^{\infty} |a_i|^2 , \quad \text{where } \mathbf{a} = (a_1, a_2, \ldots).$$ (2.10)

Note that the condition for a general sequence in $\mathbb{C}^\infty$ to be in the subspace l_2 is precisely that its norm as defined in Eq. (2.10) be finite.

Exercise. Show that the sequence of vectors $(1, 0, 0, \ldots)$, $(0, \frac{1}{2}, 0, \ldots)$, $(0, 0, \frac{1}{3}, \ldots)$ converges strongly in l_2 to the null vector.

(3) Let V be the vector space $C^0([a, b], \mathbb{C})$, i.e., the space of all continuous functions from the closed interval $[a, b]$ of the real line into the complex numbers.

Then a norm on V can be defined by

$$\|f\| := \max_{x \text{ in } [a, b]} |f(x)|. \tag{2.11}$$

Another possible norm on this vector space of functions is

$$\|f\|^2 := \int_a^b |f(x)|^2 \, dx. \tag{2.12}$$

In a precise sense (which I will not discuss here), these two norms on V are 'inequivalent' and yield different conditions for when a sequence of vectors is, or is not, strongly convergent.

In general, if a sequence of vectors $\mathbf{v}^1, \mathbf{v}^2, \ldots$ converges strongly to a vector $\mathbf{v}$ in a normed vector space V then we might expect intuitively that the sequence of real numbers $\|\mathbf{v}^1\|, \|\mathbf{v}^2\|, \ldots$ should converge in the usual sense to the real number $\|\mathbf{v}\|$. Note that this is *not* the definition of convergence, so there is a genuine result needing to be proved.

Theorem.

If the sequence of vectors $\mathbf{v}^1, \mathbf{v}^2, \ldots$ converges strongly to $\mathbf{v}$, then the sequence of real numbers $\|\mathbf{v}^1\|, \|\mathbf{v}^2\|, \ldots$ converges to $\|\mathbf{v}\|$.

Proof.

$\|\mathbf{v}\| = \|(\mathbf{v} - \mathbf{v}^N) + \mathbf{v}^N\| \le \|\mathbf{v} - \mathbf{v}^N\| + \|\mathbf{v}^N\|$ (by triangle inequality). Therefore, $\|\mathbf{v}\| - \|\mathbf{v}^N\| \le \|\mathbf{v} - \mathbf{v}^N\|$. Similarly, $\|\mathbf{v}^N\| = \|(\mathbf{v}^N - \mathbf{v}) + \mathbf{v}\| \le \|\mathbf{v}^N - \mathbf{v}\| + \|\mathbf{v}\|$ and hence $\|\mathbf{v}^N\| - \|\mathbf{v}\| \le$

$\| \mathbf{v} - \mathbf{v}^N \|$. Thus, $0 \leq |\; \| \mathbf{v}^N \| - \| \mathbf{v} \|\; | \leq \| \mathbf{v} - \mathbf{v}^N \|$, and taking the limit as $N \to \infty$, we get the result.

QED.

But suppose we are given a sequence of vectors $\mathbf{v}^1$, $\mathbf{v}^2$, ... in a normed vector space; how can we *tell* that it converges to anything at all? It hardly seems feasible to take each vector $\mathbf{v}$ in V in turn and try out the strong convergence condition on the offchance that the sequence happens to converge to it. In order to show that the sequence did *not* converge, it would be necessary to do this for every vector in V — which could be an awful lot of vectors! What is needed is a condition that can be applied directly to the sequence $\mathbf{v}^1$, $\mathbf{v}^2$, ... and which, if satisfied, guarantees that at least the sequence converges to *something*. Once we know that, we can start worrying about how to compute what the limit vector actually is.

In the analogous case of complex numbers, the desired condition is contained in the famous Cauchy condition:

"If μ_1, μ_2, ... is a sequence of complex numbers satisfying the condition that:

For any $\varepsilon > 0$, there exists $n_0(\varepsilon)$ such that $n > n_0$ and $m > n_0$ implies that $|\mu_n - \mu_m| < \varepsilon$

then the series converges to some complex number μ."

This motivates the definition of a "Cauchy convergent sequence of vectors".

Definition.

A sequence of vectors $\mathbf{v}^1$, $\mathbf{v}^2$, ... in a normed vector space V, is a *Cauchy sequence of vectors* if:

For any $\varepsilon > 0$, there exists $n_0(\varepsilon)$ such that $n > n_0$ and $m > n_0$ implies that $\| \mathbf{v}^n - \mathbf{v}^m \| < \varepsilon$.

If Cauchy convergence is to be of any use at all as a test of strong convergence, it is necessary that, at the very least, a sequence that is known to be strongly convergent should also be a Cauchy sequence. Fortunately this is true.

Theorem.

If the sequence v^1, v^2, ... converges strongly to a vector v in V, then it is a Cauchy sequence.

Proof.

$\| v^n - v^m \| = \| (v^n - v) + (v - v^m) \| \leq \| v^n - v \| + \| v - v^m \|$ by triangle inequality. But $v^n \to v$ and hence, for any $\varepsilon > 0$ there exists $n_0(\varepsilon)$ such that $n > n_0$ implies that $\| v - v^n \| < \varepsilon/2$.

But then, if $n > n_0$ and $m > n_0$, we have $\| v^n - v^m \| < \varepsilon/2 + \varepsilon/2 = \varepsilon$. Which proves the theorem.

QED.

It is comforting to know that this theorem is true, but we are really interested in the converse. Namely, is a Cauchy convergent sequence of vectors in a normed vector space V necessarily strongly convergent? For a sequence of complex numbers, the answer is "yes" but, unfortunately, this is not the case in general. Spaces for which the converse *is* true are of great importance and merit a special name.

Definition.

A normed vector space is said to be *complete* (or to be a *Banach space*) if every Cauchy convergent sequence is strongly convergent.

Spaces of this type are the only sort in which it is possible to carry out analysis in a way that resembles the familiar treatment of real or complex numbers, and therefore it is very important to know when a normed space is, or is not, complete.

Examples.

(1) As already mentioned, the complex numbers $\mathbb{C}$, regarded as a one-dimensional normed vector space, are complete and this can be used to show that the same is true for any finite-dimensional normed vector space.

(2) The space l_2 with the norm in Eq. (2.10) is complete, as is the space C^0 ([a, b], $\mathbb{C}$) with the norm defined in Eq. (2.11). Both these spaces are of course infinite-dimensional.

(3) However, the space C^0 ([a, b], $\mathbb{C}$) is *not* complete with respect to the norm defined by Eq. (2.12). This is because there exist plenty of sequences of functions $f_1, f_2, \ldots$ for which

$$\int_a^b |f_n(x) - f_m(x)|^2 \, dx$$

is a Cauchy sequence of real numbers, but which do not converge to a function on [a, b] that is continuous.

(4) Another incomplete space is $\mathscr{L}^2(\mathbb{R}^n)$ defined in Eq. (1.13) and equipped with the norm

$$\|f\|^2 := \int_{-\infty}^{\infty} \ldots \int_{-\infty}^{\infty} |f(x)|^2 \, dx^1 \, dx^2 \ldots dx^n. \quad (2.13)$$

It is by no means a trivial matter to show whether or not a specific infinite dimensional space is complete, and we will not pursue the matter further except to remark that in many ways, an incomplete space resembles the rational numbers which are certainly incomplete but which can be "completed" to form the real numbers. Similarly, an incomplete space can be made complete by "filling in the gaps", rather as the real numbers interpolate between the rationals. It is by no means obvious, however, that the elements added to an incomplete vector space necessarily resemble the type of mathematical object (eg., functions) in the space that they are completing.

2.3. SCALAR PRODUCTS

One of the most effective ways of generating norms on a vector space is via the more general concept of a "scalar product". This can be motivated by the observation that, in ordinary 3-dimensional vector calculus, the length (i.e., the norm) of a vector **v** is defined to be $(\mathbf{v} \cdot \mathbf{v})^{1/2}$ in terms of the dot product of the vector with itself. In a general vector space, a scalar product is defined to be an analogue of this 3-dimensional dot product, suitably amended to allow for the fact that we are dealing with scalar multiplication by complex, rather than real, numbers. With this proviso, the following definition mimics rather closely the crucial properties of $\mathbf{v} \cdot \mathbf{w}$.

Definition.

A *scalar product* (or *inner product*) on a complex vector space V is an assignment to each pair of vectors $\mathbf{v}$ and $\mathbf{w}$ in V of a complex number $\langle \mathbf{v}, \mathbf{w} \rangle$ satisfying the following conditions

(a) $\langle \mathbf{v}, (\mu_1 \mathbf{w}_1 + \mu_2 \mathbf{w}_2) \rangle = \mu_1 \langle \mathbf{v}, \mathbf{w}_1 \rangle + \mu_2 \langle \mathbf{v}, \mathbf{w}_2 \rangle ,$ (3.1)

(b) $\langle \mathbf{v}, \mathbf{w} \rangle^* = \langle \mathbf{w}, \mathbf{v} \rangle$ (3.2)

(c) $\langle \mathbf{v}, \mathbf{v} \rangle \geq 0$ with $\langle \mathbf{v}, \mathbf{v} \rangle = 0$ only if $\mathbf{v} = 0$. (3.3)

Note. (1) Conditions (3.1) and (3.2) imply that

$$\langle (\mu_1 \mathbf{v}_1 + \mu_2 \mathbf{v}_2), \mathbf{w} \rangle = \mu_1^* \langle \mathbf{v}_1, \mathbf{w} \rangle + \mu_2^* \langle \mathbf{v}_2, \mathbf{w} \rangle . \quad (3.4)$$

It is important to note that a complex number "comes out of the left" of the inner product with a complex conjugation whereas it emerges from the right of the bracket as it is. A failure to remember this is a common cause of difficulty when attempting to solve problems involving the scalar product.

(2) The definition of a scalar product on a real vector space is the same as on a complex space except that μ_1, μ_2 are now real numbers and there is no complex conjugation in Eq. (3.2) since $\langle \mathbf{v}, \mathbf{w} \rangle$ is now a real number for all vectors $\mathbf{v}$ and $\mathbf{w}$.

Examples.

(1) On the complex vector space $\mathbb{C}^n$, a scalar product is

$$\langle \mathbf{a}, \mathbf{b} \rangle := \sum_{i=1}^{n} a_i^* b_i , \qquad (3.5)$$

where $\mathbf{a} = (a_1, a_2, \ldots, a_n)$ and $\mathbf{b} = (b_1, b_2, \ldots, b_n)$.

Note that this is a direct analogue of the dot product in ordinary 3-dimensional vector calculus which is $\mathbf{v} \cdot \mathbf{w} := v_x w_x + v_y w_y + v_z w_z$.

(2) On the space l_2 of square-summable sequences, an obvious choice for a scalar product in analogy with Eq. (3.5), is

$$\langle \mathbf{a}, \mathbf{b} \rangle := \sum_{i=1}^{\infty} a_i^* b_i , \qquad (3.6)$$

where $\mathbf{a}$ and $\mathbf{b}$ are now the infinite sequences $\mathbf{a} = (a_1, a_2, \ldots)$ and $\mathbf{b} = (b_1, b_2, \ldots)$. However, in order that Eq. (3.6) be a sensible definition, it is necessary that the conditions

$$\sum_{i=1}^{\infty} |a_i|^2 < \infty \quad \text{and} \quad \sum_{i=1}^{\infty} |b_i|^2 < \infty \qquad (3.7)$$

should be sufficient to guarantee that the sum in Eq. (3.5) converges absolutely. That this is indeed the case, follows from the Schwarz inequality to be proved below.

(3) On the space of continuous functions $C^0([a, b], \mathbb{C})$ from the interval $[a, b]$ into the $\mathbb{C}$-numbers [to which we gave a norm in Eq. (2.12)], we can define a scalar product by

$$\langle \mathbf{f}, \mathbf{g} \rangle := \int_a^b f(x)^* \, g(x) \, dx . \qquad (3.8)$$

Similarly, on the space $\mathscr{L}^2(\mathbb{R}^n)$ [to which we gave a norm in Eq. (2.13)], we can define a scalar product by

$$\langle \mathbf{f}, \mathbf{g} \rangle := \int_{-\infty}^{\infty} \ldots \int_{-\infty}^{\infty} f(x)^* g(x) \, dx^1 \, dx^2 \ldots dx^n. \quad (3.9)$$

In both these cases, the convergence of the integral defining the scalar product follows, via the Schwarz inequality, from the square-integrability of the functions.

Note. Equation (3.9) is precisely the 'overlap function' used in elementary wave mechanics. This is a special example of the very general situation that the states of a quantum mechanical system are described by vectors in a complex vector space which is equipped with a scalar product. The fundamental probabilistic interpretation of the quantum theory is expressed directly using this scalar product. This is a very important idea and one to which we will return later.

It should be noted that, in all the examples given above, we had already defined a norm on the spaces concerned in Sec. 2.2 and that this norm is related to the scalar product by the equation $\|\mathbf{v}\|^2 = \langle v, v \rangle$, just as we define the length of the 3-vector $\mathbf{v}$ as $(\mathbf{v} \cdot \mathbf{v})^{1/2}$. This suggests rather strongly that, given a scalar product on a complex vector space V, a norm can always be constructed in this way from the scalar product. To see that this is indeed the case we need to prove the famous Schwarz inequality.

Theorem (Schwarz Inequality).

Any scalar product on a complex vector space V satisfies the inequality,

$$|\langle \mathbf{v}, \mathbf{w} \rangle| \leq \langle \mathbf{v}, \mathbf{v} \rangle^{\frac{1}{2}} \langle \mathbf{w}, \mathbf{w} \rangle^{\frac{1}{2}} \quad (3.10)$$

for all vectors **v** and **w** in V. The equality holds if and only if the vectors **v** and **w** are linearly dependent.

[**Note.** In ordinary vector calculus, this inequality is $|\mathbf{v}\cdot\mathbf{w}| \leq (\mathbf{v}\cdot\mathbf{v})^{\frac{1}{2}}(\mathbf{w}\cdot\mathbf{w})^{\frac{1}{2}}$ which, if $\cos\theta$ is the cosine of the angle between **v** and **w**, is merely the assertion that $|\cos\theta| \leq 1$.]

Proof.

(a) If $\mathbf{v} = 0$ or $\mathbf{w} = 0$ then the theorem is trivially true. If $\mathbf{v} \neq 0$ and $\mathbf{w} \neq 0$ then, for any complex number μ, the basic conditions (3.1–3) on the scalar product give the inequality:

$$0 \leq \langle(\mathbf{v} + \mu\mathbf{w}), (\mathbf{v} + \mu\mathbf{w})\rangle = \langle\mathbf{v}, \mathbf{v}\rangle + \mu^*\mu\langle\mathbf{w}, \mathbf{w}\rangle \\ + \mu^*\langle\mathbf{w}, \mathbf{v}\rangle + \mu\langle\mathbf{v}, \mathbf{w}\rangle. \quad (3.11)$$

In particular, this is true for the (cunning!) choice of

$$\mu := -\langle\mathbf{w}, \mathbf{v}\rangle/\langle\mathbf{w}, \mathbf{w}\rangle$$

which, when substituted into the inequality, gives:

$$0 \leq \langle\mathbf{v}, \mathbf{v}\rangle + \frac{|\langle\mathbf{w}, \mathbf{v}\rangle|^2}{\langle\mathbf{w}, \mathbf{w}\rangle} - \frac{\langle\mathbf{w}, \mathbf{v}\rangle^*\langle\mathbf{w}, \mathbf{v}\rangle}{\langle\mathbf{w}, \mathbf{w}\rangle} - \frac{\langle\mathbf{w}, \mathbf{v}\rangle\langle\mathbf{v}, \mathbf{w}\rangle}{\langle\mathbf{w}, \mathbf{w}\rangle}.$$

The right hand side of this expression is

$$\langle\mathbf{v}, \mathbf{v}\rangle + \frac{|\langle\mathbf{w}, \mathbf{v}\rangle|^2}{\langle\mathbf{w}, \mathbf{w}\rangle} - \frac{2\,|\langle\mathbf{w}, \mathbf{v}\rangle|^2}{\langle\mathbf{w}, \mathbf{w}\rangle}$$

and so,

$$0 \leq \langle\mathbf{v}, \mathbf{v}\rangle - \frac{|\langle\mathbf{w}, \mathbf{v}\rangle|^2}{\langle\mathbf{w}, \mathbf{w}\rangle}, \text{ i.e., } |\langle\mathbf{w}, \mathbf{v}\rangle|^2 \leq \langle\mathbf{v}, \mathbf{v}\rangle\langle\mathbf{w}, \mathbf{w}\rangle.$$

(b) If $\mathbf{v}$ and $\mathbf{w}$ are linearly dependent, then substitution of $\mathbf{v} = \mu\mathbf{w}$ into (3.10) shows at once that the equality holds.

Conversely, if the equality sign holds in the final expression obtained in the argument above, then it must hold at all stages in the argument. In particular, it must hold in Eq. (3.11) and hence, by condition (3.3), we must have $\mathbf{v} + \mu\mathbf{w} = 0$, which means that $\mathbf{v}$ and $\mathbf{w}$ are linearly dependent.

QED.

Corollary.

If $\langle\ ,\ \rangle$ is a scalar product on a vector space V, then a norm can be defined on V by

$$\|\mathbf{v}\| := \langle \mathbf{v}, \mathbf{v} \rangle^{\frac{1}{2}}. \tag{3.12}$$

Proof.

Conditions (2.5) and (2.6) on a norm follow at once from the defining conditions of a scalar product.

To prove the triangle inequality (2.4) we proceed as follows

$$\begin{aligned}
\langle (\mathbf{v} + \mathbf{w}), (\mathbf{v} + \mathbf{w}) \rangle &= \langle \mathbf{v}, \mathbf{v} \rangle + \langle \mathbf{w}, \mathbf{w} \rangle + \langle \mathbf{v}, \mathbf{w} \rangle + \langle \mathbf{w}, \mathbf{v} \rangle \\
&= \langle \mathbf{v}, \mathbf{v} \rangle + \langle \mathbf{w}, \mathbf{w} \rangle + 2\,\mathrm{Re}\,\langle \mathbf{v}, \mathbf{w} \rangle \\
&\leq \langle \mathbf{v}, \mathbf{v} \rangle + \langle \mathbf{w}, \mathbf{w} \rangle + 2\,|\langle \mathbf{v}, \mathbf{w} \rangle| \\
&\leq \langle \mathbf{v}, \mathbf{v} \rangle + \langle \mathbf{w}, \mathbf{w} \rangle + 2\langle \mathbf{v}, \mathbf{v} \rangle^{\frac{1}{2}}\langle \mathbf{w}, \mathbf{w} \rangle^{\frac{1}{2}}
\end{aligned}$$
from the theorem.
$$= (\langle \mathbf{v}, \mathbf{v} \rangle^{\frac{1}{2}} + \langle \mathbf{w}, \mathbf{w} \rangle^{\frac{1}{2}})^2.$$

This asserts precisely that $\|\mathbf{v} + \mathbf{w}\| \leq \|\mathbf{v}\| + \|\mathbf{w}\|$.   QED.

Note. (1) Just as in the case of a norm, it is important to distinguish those vector spaces that are complete. In particular, a complete normed space in which the norm is obtained as above from a scalar product is known as *Hilbert space*. One of the basic axioms of conventional quantum theory is that the states of the quantum system are represented by vectors in a Hilbert space.

(2) Examples of Hilbert spaces are the finite-dimensional space $\mathbb{C}^n$ and the infinite-dimensional space l_2. As it stands, the space $\mathscr{L}^2(\mathbb{R}^n)$ is *not* complete. However, it can be completed successfully by adding in certain limit functions and using Lebesgue, rather than Riemann, integration in the definition of the inner product. In employing wavefunctions in elementary quantum theory, it is in fact assumed implicitly that this completion has been performed. To distinguish between the incomplete space $\mathscr{L}^2(\mathbb{R}^n)$ and its completion, it is customary to denote the latter by $L^2(\mathbb{R}^n)$.

(3) It is not the case that *all* norms can be obtained from a scalar product via the mechanism above. For example, there is *no* scalar product on the space $C^0([a, b], \mathbb{C})$ that will yield the norm defined in Eq. (2.11).

Much of the geometrical language used in connection with the dot product in elementary 3-dimensional vector space theory is carried across to the general case of a scalar product on a complex vector space.

Definitions.

(a) Let V be a vector space with an inner product $\langle \, , \rangle$. Then two vectors **v** and **w** are said to be *orthogonal* if

$$\langle \mathbf{v}, \mathbf{w} \rangle = 0 \, .$$

(b) A vector **v** is said to be *normalized* if $\| \mathbf{v} \| := \langle \mathbf{v}, \mathbf{v} \rangle^{\frac{1}{2}} = 1$.

(c) A pair of vectors **v**, **w** is said to be an *orthonormal pair* if $\langle \mathbf{v}, \mathbf{w} \rangle = 0$ and $\| \mathbf{v} \| = \| \mathbf{w} \| = 1$.

(d) An *orthonormal subset* of vectors in V is a set of vectors in which every vector is normalized and every pair of vectors is orthogonal to each other.

(e) In a finite-dimensional vector space, an *orthonormal basis* is a basis set $\{\mathbf{e}^1, \mathbf{e}^2, \ldots, \mathbf{e}^n\}$ [where $n = \dim(V)$] which is an orthonormal subset of V, i.e.,

$$\langle \mathbf{e}^i, \mathbf{e}^j \rangle = \delta^{ij} \quad \text{for } i, j = 1, \ldots, n. \tag{3.13}$$

Note. (1) Every finite-dimensional vector space has an orthonormal basis. (Exercise: Prove this by starting with any basis set and take suitable linear combinations to get an orthonormal set. This is known as the Gram-Schmidt procedure.)

(2) Let $\{\mathbf{e}^1, \mathbf{e}^2, \ldots, \mathbf{e}^n\}$ be an orthonormal basis for V and consider the expansion of any vector in terms of this basis [cf. Eq. (1.15)]

$$\mathbf{v} = \sum_{i=1}^{n} v_i \mathbf{e}^i. \tag{3.14}$$

One of the great advantages of using an orthonormal basis is that it is possible to compute *explicitly* the expansion coefficients knowing only the vector **v** and the basis set in use. To see this, let us take the scalar product of both sides of Eq. (3.14) with a particular basis element, say $\mathbf{e}^j$

$$\langle \mathbf{e}^j, \mathbf{v} \rangle = \langle \mathbf{e}^j, \sum_{i=1}^{n} v_i \, \mathbf{e}^i \rangle \tag{3.15}$$

$$= \sum_{i=1}^{n} \langle \mathbf{e}^j, v_i \, \mathbf{e}^i \rangle \tag{3.16}$$

$$= \sum_{i=1}^{n} v_i \langle \mathbf{e}^j, \mathbf{e}^i \rangle = \sum_{i=1}^{n} v_i \, \delta^{ij} = v_j.$$

Thus we see that the coefficient v_j is given in terms of the vector $\mathbf{v}$ and the basis set, as

$$\boxed{v_j = \langle \mathbf{e}^j, \mathbf{v} \rangle.} \tag{3.17}$$

This result is of considerable importance from both a practical and a theoretical point of view.

Note that Eq. (3.17) means that we can write as an identity

$$\mathbf{v} = \sum_{i=1}^{n} \langle \mathbf{e}^i, \mathbf{v} \rangle \, \mathbf{e}^i \tag{3.18}$$

and similarly,

$$\| \mathbf{v} \|^2 = \left\langle \left(\sum_{i=1}^{n} \langle \mathbf{e}^i, \mathbf{v} \rangle \mathbf{e}^i \right), \mathbf{v} \right\rangle = \sum_{i=1}^{n} |\langle \mathbf{e}^i, \mathbf{v} \rangle|^2 . \tag{3.19}$$

A special case of this last result occurs in ordinary 3-dimensional vector calculus when we expand a vector $\mathbf{v}$ with respect to the orthonormal basis set $\{\mathbf{i}, \mathbf{j}, \mathbf{k}\}$ as $\mathbf{v} = v_x \mathbf{i} + v_y \mathbf{j} + v_z \mathbf{k}$. The expansion coefficients are given by the appropriate analogue of Eq. (3.17), viz. $v_x = \mathbf{i} \cdot \mathbf{v}$ etc. and the analogue of (3.19) is the well-known expression

$$\mathbf{v} \cdot \mathbf{v} = v_x^2 + v_y^2 + v_z^2.$$

We come now to the very important question concerning the extent to which the ideas of orthonormal bases and the associated expansion theorems can be generalized to include the case when the vector space is infinite-dimensional. One might be tempted to simply try writing the analogue of, for example, Eq. (3.18) with the upper limit of the sum equal to infinity rather than the finite integer n. However, in order to justify such a step, we must first consider the following:

(1) As we have already discussed in Sec. 2.2, the infinite sum in Eq. (3.18) must be interpreted as the strong limit of the partial sums and, for the same reasons as before, we shall therefore require that the vector space be complete, i.e., it must be a Hilbert space.

(2) In deriving the crucial result (3.17), we had to take the summation sign from inside the scalar product $\langle , \rangle$ in Eq. (3.15) and place it outside the scalar product as in Eq. (3.16). For a finite sum, this is justified by the linearity condition (3.1) in the definition of a scalar product. However, this does *not* automatically justify the interchange in the infinite-dimensional case and a theorem needs to be proved.

(3) In writing an equation like (3.18) with an infinite upper limit, we are evidently assuming that, in some sense, the Hilbert space is "countably infinite". But need this always be the case? Perhaps the situation could arise in which the dimension of the space is a higher-order infinity (like the number of real numbers, rather than the number of integers) and then one might anticipate having a basis set labelled by a real number, and with the sum in Eq. (3.18) replaced by an integral.

There are indeed occasions in quantum theory where spaces of this type do occur. But they can usually be avoided and we will certainly do so in this course. The technical version of the intuitive idea that a space has a countably infinite dimension is contained in the following definition.

Definitions.

(a) A Hilbert space is said to be *separable* if there is some subset S of vectors of V such that

   (i) Every vector in V can be written as a finite linear combination of elements in S, or as the strong limit of such sums.

   (ii) The number of elements in S is finite or countably infinite.

(b) If the elements of S are in addition linearly independent of each other then S is said to be a *basis* for V.

Note. (1) As in the finite-dimensional case, it is always possible to arrange things so that S is an orthonormal set of vectors.

   (2) A finite-dimensional Hilbert space is separable by definition.

   (3) A good example of a separable, infinite-dimensional, Hilbert space is l_2. This has a countable orthonormal basis made up of the vectors:

$$\{(1, 0, 0, \ldots), (0, 1, 0, \ldots), (0, 0, 1, \ldots), \ldots\}.$$

A less obvious (but very important) example is the vector space $L^2([0, 1])$ which is the completion of the space $C^\infty([0, 1], \mathbb{C})$ with respect to the scalar product in Eq. (3.8). An orthonormal basis set for this space is the set of functions

$$\{1, 2^{\frac{1}{2}} \sin 2\pi x, 2^{\frac{1}{2}} \cos 2\pi x, 2^{\frac{1}{2}} \sin 4\pi x, 2^{\frac{1}{2}} \cos 4\pi x, \ldots\}.$$
$$(3.20)$$

In other words, when we expand a function as a Fourier sum, we are regarding the function as a vector in $L^2([0, 1])$ and expanding it in terms of the particular basis set given in Eq. (3.20). This means, in particular, that the Fourier sum converges in the sense that its partial sums converge strongly with respect to the norm derived from Eq. (3.8). Note that this does *not* imply that the infinite sum of functions converges pointwise for each value of x in the interval [0, 1], and indeed, most Fourier sums do *not* converge in this way. The strong norm convergence is often referred to as "convergence in the mean" in this particular context.

A critical element in the justification of the expansion theorems when V is a separable Hilbert space is the following lemma. This shows how, in a well-defined sense, the scalar product $\langle v, w \rangle$ is a "continuous function" of the vectors v and w.

Lemma.

Let $v^n \to v$ be a strongly convergent sequence of vectors in a Hilbert space $\mathcal{H}$. Then, for all vectors w in $\mathcal{H}$,

$$\lim_{n \to \infty} \langle w, v^n \rangle = \langle w, v \rangle. \qquad (3.21)$$

Proof.

$$0 \le |\langle w, v^n \rangle - \langle w, v \rangle| = |\langle w, (v^n - v) \rangle|$$

$$\le \| w \| \, \| v^n - v \| \qquad \text{by the Schwarz inequality.}$$

Then, taking the limit as $n \to \infty$ on both sides of this second inequality we find that, since $\lim_{n \to \infty} \| v^n - v \| = 0$ (as v^n converges strongly to v)

$$\lim_{n \to \infty} |\langle w, v^n \rangle - \langle w, v \rangle| = 0$$

which proves the result.

QED.

Note. It is trivial to prove from this that we also have,

$$\lim_{n\to\infty} \langle \mathbf{v}^n, \mathbf{w} \rangle = \langle \mathbf{v}, \mathbf{w} \rangle. \qquad (3.22)$$

Another key result is the following.

Lemma (Bessel's Inequality).

Let $\mathcal{H}$ be a separable Hilbert space with an orthonormal basis $\{\mathbf{e}^1, \mathbf{e}^2, \ldots\}$. Then, for any positive integer N,

$$\sum_{i=1}^{N} |\langle \mathbf{e}^i, \mathbf{v} \rangle|^2 \le \|\mathbf{v}\|^2 \quad \text{for all vectors } \mathbf{v} \text{ in } \mathcal{H}. \qquad (3.23)$$

Proof.

For a fixed value of N, let us try to approximate a given vector $\mathbf{v}$ as closely as possible by a linear combination of $\{\mathbf{e}^1, \mathbf{e}^2, \ldots, \mathbf{e}^N\}$. More precisely, we vary the coefficients μ_i so as to minimize the value of the expression $\|\mathbf{v} - \sum_{i=1}^{N} \mu_i \mathbf{e}^i\|^2$. Now,

$$\left\| \mathbf{v} - \sum_{i=1}^{N} \mu_i \mathbf{e}^i \right\|^2 = \left\langle \mathbf{v} - \sum_{i=1}^{N} \mu_i \mathbf{e}^i, \mathbf{v} - \sum_{j=1}^{N} \mu_j \mathbf{e}^j \right\rangle$$

$$= \langle \mathbf{v}, \mathbf{v} \rangle - \sum_{i=1}^{N} \mu_i^* \langle \mathbf{e}^i, \mathbf{v} \rangle - \sum_{j=1}^{N} \mu_j \langle \mathbf{v}, \mathbf{e}^j \rangle + \sum_{i=1}^{N} \mu_i^* \mu_i$$

$$= \|\mathbf{v}\|^2 + \sum_{i=1}^{N} |\langle \mathbf{e}^i, \mathbf{v} \rangle - \mu_i|^2 - \sum_{i=1}^{N} |\langle \mathbf{e}^i, \mathbf{v} \rangle|^2$$

and it is clear that the minimum will be reached when $\mu_i = \langle e^i, v \rangle$. For this value of μ_i we then have

$$\left\| v - \sum_{i=1}^{N} \mu_i\, e^i \right\|^2 = \| v \|^2 - \sum_{i=1}^{N} |\langle e^i, v \rangle|^2$$

and hence,

$$\| v \|^2 \geq \sum_{i=1}^{N} |\langle e^i, v \rangle|^2 . \qquad \text{QED.}$$

Now we come to the central expansion theorem for an infinite-dimensional separable, Hilbert space.

Theorem.

Let $\mathcal{H}$ be an infinite-dimensional, separable, Hilbert space with an orthonormal basis set $\{e^1, e^2, \ldots\}$. Then any vector v in $\mathcal{H}$ can be expanded as

$$v = \sum_{i=1}^{\infty} \langle e^i, v \rangle\, e^i , \qquad (3.24)$$

where the infinite sum means the strong limit of the partial sums

$$S^N := \sum_{i=1}^{N} \langle e^i, v \rangle\, e^i .$$

Proof.

(a) From Bessel's inequality we have, for all N, $\sum_{i=1}^{N} |\langle e^i, v \rangle|^2 \leq \| v \|^2$ and so, taking the limit as $N \to \infty$ we find

$$\sum_{i=1}^{\infty} |\langle \mathbf{e}^i, \mathbf{v}\rangle|^2 \leq \| \mathbf{v} \|^2 .$$

Thus the left hand side of this inequality is a convergent sum of real numbers and hence its partial sums must be a Cauchy sequence. Therefore, as m and n tend to infinity, we have

$$\left\| \sum_{i=m}^{n} \langle \mathbf{e}^i, \mathbf{v}\rangle \mathbf{e}^i \right\|^2 = \sum_{i=m}^{n} |\langle \mathbf{e}^i, \mathbf{v}\rangle|^2 \to 0$$

which means in turn that the partial sums $S^N := \Sigma_{i=1}^{N}\langle \mathbf{e}^i, \mathbf{v}\rangle \mathbf{e}^i$ form a Cauchy sequence of vectors in $\mathcal{H}$. But $\mathcal{H}$ is complete and hence S^N must converge to some vector $\mathbf{w} := \Sigma_{i=1}^{\infty}\langle \mathbf{e}^i, \mathbf{v}\rangle \mathbf{e}^i$.

(b) Given that S^N converges strongly to $\mathbf{w} := \Sigma_{i=1}^{\infty}\langle \mathbf{e}^i, \mathbf{v}\rangle \mathbf{e}^i$, does $\mathbf{w} = \mathbf{v}$? To answer this, take a general basis vector $\mathbf{e}^j$ and study $\langle \mathbf{e}^j, \mathbf{w}\rangle$

$$\langle \mathbf{e}^j, \mathbf{w}\rangle = \left\langle \mathbf{e}^j, \lim_{N\to\infty} \sum_{i=1}^{N}\langle \mathbf{e}^i, \mathbf{v}\rangle \mathbf{e}^i \right\rangle$$

$$= \lim_{N\to\infty} \left\langle \mathbf{e}^j, \sum_{i=1}^{N}\langle \mathbf{e}^i, \mathbf{v}\rangle \mathbf{e}^i \right\rangle \quad \text{from the Lemma}$$

$$= \langle \mathbf{e}^j, \mathbf{v}\rangle .$$

Thus, for all basis vectors $\mathbf{e}^1, \mathbf{e}^2, \ldots$, we have

$$\langle \mathbf{e}^j, (\mathbf{v} - \mathbf{w})\rangle = 0 . \tag{3.25}$$

But $\{\mathbf{e}^1, \mathbf{e}^2, \ldots\}$ is a basis set and hence there must exist complex numbers $\mu_1, \mu_2, \ldots$ such that $(\mathbf{v} - \mathbf{w}) = \Sigma_{i=1}^{\infty}\mu_i \mathbf{e}^i$. And then,

$$\langle (\mathbf{v} - \mathbf{w}), (\mathbf{v} - \mathbf{w}) \rangle = \left\langle \lim_{N \to \infty} \sum_{i=1}^{N} \mu_i \, \mathbf{e}^i , (\mathbf{v} - \mathbf{w}) \right\rangle$$

$$= \lim_{N \to \infty} \sum_{i=1}^{N} \langle \mu_i \mathbf{e}^i, (\mathbf{v} - \mathbf{w}) \rangle$$

and this vanishes from Eq. (3.25). Thus $\| \mathbf{v} - \mathbf{w} \|^2 = 0$ and hence it follows that $\mathbf{v} = \mathbf{w}$, which proves Eq. (3.24).

QED.

Corollary.

(a) $\| \mathbf{v} \|^2 = \sum_{i=1}^{\infty} |\langle \mathbf{e}^i, \mathbf{v} \rangle|^2$ (Parseval's formula)   (3.26)

(b) $\langle \mathbf{v}, \mathbf{w} \rangle = \sum_{i=1}^{\infty} \langle \mathbf{v}, \mathbf{e}^i \rangle \langle \mathbf{e}^i, \mathbf{w} \rangle .$ (3.27)

Exercise. Proof the corollary.

In a finite-dimensional vector space V, the statement that a subset S of vectors is a basis for V is equivalent to the assertion that $V = [S]$, where $[S]$ — the span of S — is defined to be the set of all finite linear combinations of vectors in S. Evidently, this concept needs generalizing somewhat in the case that V is an infinite-dimensional Hilbert space. Indeed, if $S := \{ \mathbf{e}^1, \mathbf{e}^2, \ldots \}$ is a basis set for a Hilbert space $\mathcal{H}$ in the sense we have just been discussing, then $[S]$ is certainly a linear subspace of $\mathcal{H}$ but it is also strictly smaller than $\mathcal{H}$ since it does not contain the strong limit of the finite sums of basis vectors.

This situation can arise for a general linear subspace W of an infinite-dimensional Hilbert space $\mathcal{H}$. Viz. there may exist sequences of vectors $\mathbf{v}^1, \mathbf{v}^2, \ldots$ such that each $\mathbf{v}^N$ lies in W and such that $\mathbf{v}^N$ converges strongly to some vector $\mathbf{v}$ in $\mathcal{H}$ and yet $\mathbf{v}$ does not lie in the linear subspace W. This in turn means that, although W "inherits" a structure of a vector space in its own right as a subspace of $\mathscr{H}$, it is not a *Hilbert* space since it is not complete. This motivates the definition.

Definitions.

(a) The *closure* $\bar{T}$ of a set of vectors T in $\mathscr{H}$ is the union of T and the limits in $\mathscr{H}$ of all strongly convergent sequences of vectors in T.

(b) A linear subspace W of $\mathscr{H}$ is *closed* if $W = \bar{W}$, i.e., if it contains the limits of all strongly convergent sequences of its elements.

Thus to say that a subset S of vectors in $\mathscr{H}$ is a basis for $\mathscr{H}$ means precisely that $\mathscr{H} = \overline{[S]}$. It also follows from the discussion above that if W is a closed subspace of $\mathscr{H}$ then it is itself a Hilbert space, and the same is true of the quotient vector space $\mathscr{H}/W$. This is usually a highly desirable property and for this reason most references to "subspaces" of a Hilbert space implicitly or explicitly assume that the linear subspace of interest is closed.

Example.

The *orthogonal complement* $W_\perp$ of a linear subspace W of $\mathscr{H}$ is defined to be the set of all vectors in $\mathscr{H}$ that are orthogonal to W:

$$W_\perp := \{\mathbf{v} \text{ in } \mathscr{H} \text{ such that } \langle \mathbf{v}, \mathbf{w} \rangle = 0 \text{ for all } \mathbf{w} \text{ in } W\} .$$

Then, if $\mathbf{v}^n \to \mathbf{v}$ is a strongly convergent sequence of vectors in $W_\perp$ we have, for all $\mathbf{w}$ in W, $\langle \mathbf{v}, \mathbf{w} \rangle = \langle \lim_{n \to \infty} \mathbf{v}^n, \mathbf{w} \rangle = \lim_{n \to \infty} \langle \mathbf{v}^n, \mathbf{w} \rangle$ [Eq. (3.22)] $= 0$. Thus $\mathbf{v}$ belongs to $W_\perp$, i.e., $W_\perp$ is a closed linear subspace of $\mathscr{H}$ irrespective of whether W is closed or not.

Another important geometrical concept that arises in this context is that of the projection of a vector in $\mathcal{H}$ onto a closed subspace W.

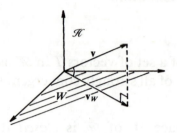

If $\mathcal{H}$ is a *finite*-dimensional Hilbert space with an orthonormal basis set $\{e^1, e^2, \ldots, e^n\}$ and if W is an m-dimensional linear subspace of $\mathcal{H}$, then we can write $W = [S]$, where S is the subset $\{e^{i_1}, e^{i_2}, \ldots, e^{i_m}\}$ of basis vectors that lie in W. Then the *projection* of a vector v in $\mathcal{H}$ onto the subspace W is the vector v_W in W defined by

$$v_W := \sum_{a=1}^{m} \langle e^{i_a}, v \rangle \, e^{i_a}, \tag{3.28}$$

where $\langle e^i, v \rangle$ are, of course, the expansion coefficients of v with respect to the basis set $\{e^1, e^2, \ldots, e^n\}$ in $\mathcal{H}$.

When $\mathcal{H}$ is infinite-dimensional, we can proceed in more or less the same way provided that W is a closed subspace of $\mathcal{H}$. If $\{w^1, w^2, \ldots\}$ is an orthonormal basis set for W [eg. a subset of a basis set $\{e^1, e^2, \ldots\}$ for $\mathcal{H}$] then the *projection* of v in $\mathcal{H}$ onto W is defined as

$$v_W := \sum_{a=1}^{\infty} \langle w^a, v \rangle \, w^a \tag{3.29}$$

which exists and lies in W because W is complete.

If we define

$$\mathbf{v}_{W_\perp} := \mathbf{v} - \mathbf{v}_W \tag{3.30}$$

then $\langle \mathbf{v}_{W_\perp}, \mathbf{w}^a \rangle = 0$ for all of the orthonormal basis vectors $\mathbf{w}^a$ in W and hence $\mathbf{v}_{W_\perp}$ belongs to $W_\perp$; in fact, it is simply the projection of $\mathbf{v}$ onto the closed subspace $W_\perp$.

Thus we have obtained a decomposition of any vector $\mathbf{v}$ in $\mathscr{H}$ as

$$\mathbf{v} = \mathbf{v}_W + \mathbf{v}_{W_\perp} \tag{3.31}$$

with respect to any closed subspace W of $\mathscr{H}$. A decomposition such as this is necessarily unique (Exercise) and the situation is expressed by saying that $\mathscr{H}$ is the *direct sum* $W \oplus W_\perp$ of the closed subspaces W and $W_\perp$. This is an important concept to which we shall return later.

2.4. LINEAR OPERATORS

Linear operators are an important ingredient in the mathematical apparatus of quantum theory as they are used to represent physical observables. They are also central to the general theory of group representations and therefore we will spend a little time considering some of the basic ideas before moving on to discuss their applications in these areas.

A linear map between two vector spaces was defined in Eq. (1.6) as a map that preserves the structure of the vector space on which it is defined. A 'linear operator' is nothing else but a linear map from a vector space into itself.

Definitions.

(a) A *linear operator* on a vector space V is a linear map $A : V \to V$ from V to itself. The image $A(\mathbf{v})$ of $\mathbf{v}$ is usually written as $A\mathbf{v}$ and hence the linearity requirement is

$$A(\mu_1 \mathbf{v}_1 + \mu_2 \mathbf{v}_2) = \mu_1 A\mathbf{v}_1 + \mu_2 A\mathbf{v}_2 \qquad (4.1)$$

for all complex numbers μ_1, μ_2 and all vectors $\mathbf{v}_1$, $\mathbf{v}_2$ in V.

(b) Two linear operators A and B are said to be *equal* if $A\mathbf{v} = B\mathbf{v}$ for all vectors $\mathbf{v}$ in V.

Note. (1) The set of all linear operators on a vector space V can itself be given the structure of a vector space by defining

   (i) $(A + B)\mathbf{v} := A\mathbf{v} + B\mathbf{v}$ for all $\mathbf{v}$ in V. (4.2)
   (ii) $(\mu A)\mathbf{v} := \mu(A\mathbf{v})$ for all $\mathbf{v}$ in V and μ in $\mathbb{C}$.

(2) The "product" of two operators A and B can be defined by

$$(AB)\mathbf{v} := A(B\mathbf{v}) \qquad \text{for all } \mathbf{v} \text{ in } V. \qquad (4.3)$$

This places a monoid structure on the set of all operators on V in which the unit element is just the identity operator $\mathbf{1}$ defined by $\mathbf{1v} := \mathbf{v}$ for all $\mathbf{v}$ in V.

(3) An operator A is *invertible* if there exists another operator A^{-1} with the property that

$$A^{-1}A = A A^{-1} = \mathbf{1} .$$

Such an operator is an isomorphism of V onto itself (cf. Sec. 2.1) and the set of all of these is the group $\text{Aut}(V)$ of automorphisms of V onto itself that was mentioned in Sec. 2.1. Thus $\text{Aut}(V)$ is a group that is a subset of the monoid of all operators on V.

Example.

Let V be a finite-dimensional vector space with a basis $\{\mathbf{e}^1, \mathbf{e}^2, \ldots, \mathbf{e}^n\}$. Then any vector $\mathbf{v}$ can be expanded as

$$\mathbf{v} = \sum_{i=1}^{n} v_i\, \mathbf{e}^i \tag{4.4}$$

and hence, if A is any linear operator on V, we can apply A to both sides of Eq. (4.4) and use the linearity to get

$$A\mathbf{v} = A\left(\sum_{i=1}^{n} v_i\, \mathbf{e}^i\right) = \sum_{i=1}^{n} v_i\, A\mathbf{e}^i. \tag{4.5}$$

Since $\{\mathbf{e}^1, \mathbf{e}^2, \ldots, \mathbf{e}^n\}$ is a basis set, there must exist complex numbers A_{ji}, $i, j = 1, \ldots, n$ such that

$$A\mathbf{e}^i = \sum_{j=1}^{n} \mathbf{e}^j A_{ji} \quad \text{for each } i = 1, \ldots, n. \tag{4.6}$$

Thus,

$$A\mathbf{v} = \sum_{i=1}^{n}\sum_{j=1}^{n} v_i A_{ji}\, \mathbf{e}^j = \sum_{j=1}^{n}\left(\sum_{i=1}^{n} A_{ji} v_i\right)\mathbf{e}^j.$$

It follows that we can represent the action of the linear operator A by what it does to the components $(v_1, v_2, \ldots, v_n)$ of a vector $\mathbf{v}$, by means of the matrix equation

$$\begin{pmatrix} v_1 \\ v_2 \\ \vdots \\ v_n \end{pmatrix} \rightsquigarrow \begin{pmatrix} A_{11} & A_{12} & \ldots & A_{1n} \\ A_{21} & A_{22} & \ldots & A_{2n} \\ \vdots & \vdots & & \vdots \\ A_{n1} & A_{n2} & \ldots & A_{nn} \end{pmatrix} \begin{pmatrix} v_1 \\ v_2 \\ \vdots \\ v_n \end{pmatrix}, \tag{4.7}$$

where the matrix A_{ij}, $i, j = 1, \ldots, n$ is called the *matrix version* (or *matrix representation*) of the operator A with respect to the given basis.

In terms of the isomorphism $i : V \to \mathbb{C}^n$ introduced in Sec. 2.1 and defined by $i(v) := (v_1, v_2, \ldots, v_n)$, we have the commutative diagram

$$
\begin{array}{ccc}
V & \xrightarrow{\quad i \quad} & \mathbb{C}^n \\
\Big\downarrow{\scriptstyle A} & & \Big\downarrow{\scriptstyle A_{ij}} \\
V & \xrightarrow{\quad i \quad} & \mathbb{C}^n
\end{array}
$$

where A is the original operator on the vector space V and A_{ij} refers to its matrix representative.

(4.8)

Note that this relationship between operators and matrices is one-to-one, and that every $n \times n$ square matrix induces a linear operator on V via Eq. (4.6) now regarded as a *definition* of A.

It should also be noted that the matrix representation of the product AB of two operators is just the matrix product of the matrix representations of A and B. (Exercise) It follows, in particular, that the monoid of all operators on V is isomorphic to the monoid $M(n, \mathbb{C})$ with the multiplicative structure defined in Sec. 1.1.

Similarly, an operator A is invertible as an operator if and only if the matrix representation is invertible as a matrix, i.e., if it belongs to the subgroup $GL(n, \mathbb{C})$ of the monoid $M(n, \mathbb{C})$. (Exercise)

Note. (1) If V has an inner product and if $\{e^1, e^2, \ldots, e^n\}$ is an orthonormal basis then taking the inner product of both sides of Eq. (4.6) with a particular basis element e^k, gives

$$
\langle e^k, Ae^i \rangle = \left\langle e^k, \sum_{j=1}^{n} e^j A_{ji} \right\rangle
$$

$$
= \sum_{j=1}^{n} \langle e^k, e^j A_{ji} \rangle = A_{ki} .
$$

Thus,

$$A_{ij} = \langle \mathbf{e}^i, A\mathbf{e}^j \rangle \qquad (4.9)$$

which is an important equation for the "matrix elements" of the operator A and which is often used in the formal development of quantum theory à la Dirac.

(2) In general, the matrices representing A corresponding to two different bases for V (orthonormal or otherwise) will be different, even though they represent the same abstract operator on V. This emphasises the fact that a $n \times n$ matrix has no intrinsic significance as a linear operator on V until the basis set with which it is supposed to be associated has been specified.

We would like to extend all these results to the case where V is an infinite-dimensional, separable Hilbert space and, for example, replace Eq. (4.5) with a version in which the upper limit is infinity. However, remembering that an infinite sum is defined as the strong limit of its partial sums, we see that such a step is possible only if it is true that $A(\lim_{N\to\infty} S^N) = \lim_{N\to\infty} (AS^N)$. But there is no reason why such an interchange of operator and limit should be true in general. Operators for which it *is* valid are particularly pleasant to use and have a special name.

Definitions.

(a) A linear operator on a separable Hilbert space $\mathscr{H}$ is said to be *continuous* if

$$\mathbf{v}^n \to \mathbf{v} \qquad \text{implies that } A\mathbf{v}^n \to A\mathbf{v} \qquad (4.10)$$

for all strongly convergent sequences of vectors $\mathbf{v}^n$.

(b) A linear operator on $\mathscr{H}$ is *bounded* if there exists some positive real number b such that

$$\| A\mathbf{v} \| \le b \| \mathbf{v} \| \qquad \text{for all } \mathbf{v} \text{ in } \mathscr{H}. \qquad (4.11)$$

The smallest such number b is called the *norm* of the operator and is written as $\| A \|$.

It is an important technical result that these two concepts are equivalent as shown in the following theorem.

Theorem.

An operator A is continuous if and only if it is bounded.

Proof.

(a) If A is bounded and if $\mathbf{v}^n \to \mathbf{v}$ is a strongly convergent sequence of vectors in $\mathscr{H}$, then

$$\| A\mathbf{v} - A\mathbf{v}^n \| = \| A(\mathbf{v} - \mathbf{v}^n) \| \le \| A \| \| \mathbf{v}^n - \mathbf{v} \|, \quad (4.12)$$

where the last inequality follows from the definition of the norm $\| A \|$ of the operator A. Then taking the limit as $n \to \infty$ on both sides of Eq. (4.12) we find that

$$\lim_{n \to \infty} \| A\mathbf{v} - A\mathbf{v}^n \| = 0$$

which means precisely that the sequence of vectors $A\mathbf{v}^1$, $A\mathbf{v}^2$, ... converges strongly to $A\mathbf{v}$, i.e., that A is continuous.

(b) Conversely, if A is *not* bounded, for each n there must exist some vector $\mathbf{v}^n$ such that $\| A\mathbf{v}^n \| > n \| \mathbf{v}^n \|$. Now define the sequence of vectors $\mathbf{w}^n := (n \| \mathbf{v}^n \|)^{-1} \mathbf{v}^n$. Then $\| \mathbf{w}^n \| = n^{-1}$ and hence the sequence $\mathbf{w}^n$ converges strongly to the null vector as $n \to \infty$. But $\| A\mathbf{w}^n \| > 1$ and hence $A\mathbf{w}^n$ does *not* converge to the null vector. Thus A is not continuous.

QED.

Note. (1) Every linear operator on a finite-dimensional vector space is bounded. (Exercise)

(2) A bounded linear operator on a separable Hilbert space $\mathscr{H}$ with an orthonormal basis set $\{\mathbf{e}^1, \mathbf{e}^2, \ldots\}$ can be represented by the infinite matrix A_{ij}, where

$$A\mathbf{e}^i = \sum_{j=1}^{\infty} \mathbf{e}^j A_{ji} \qquad \text{for } i = 1, 2, 3, \ldots.$$

But it is *not* the case that every infinite square matrix represents a bounded operator.

Some instructive examples of this last phenomenon are as follows.

Examples.

(1) On the Hilbert space l_2, the matrix $A :=$ cannot represent a bounded operator with respect to the basis
$\mathbf{e}^1 := (1, 0, 0, \ldots)$
$\mathbf{e}^2 := (0, 1, 0, \ldots)$ etc.
since $A\mathbf{e}^1 = \mathbf{e}^1$, $A\mathbf{e}^2 = 2\mathbf{e}^2$, $A\mathbf{e}^3 = 3\mathbf{e}^3$ and so on.

$$\begin{pmatrix} 1 & 2 & 3 & 4 & \cdots \\ 0 & 0 & 0 & 0 & \cdots \\ 0 & 0 & 0 & 0 & \cdots \\ \vdots & \vdots & \vdots & \vdots & \cdots \end{pmatrix}$$

It is important to note that, in fact, this particular matrix operator cannot even be *defined* on every vector in l_2 since it will map certain vectors out of the l_2 subspace of $\mathbb{C}^\infty$. For example, if we act with A on the vector $\begin{pmatrix} 1 \\ 1/2 \\ 1/3 \\ \vdots \end{pmatrix}$ then we get the vector

$\begin{pmatrix} 1 \\ 1 \\ 1 \\ \vdots \end{pmatrix}$ which belongs to the vector space $\mathbb{C}^\infty$ of arbitrary

sequences of complex numbers but which does not belong to l_2 since its elements are not square-summable.

This is actually a typical feature of unbounded operators as it turns out that such an operator can only be defined on some subset of vectors (known as the *domain* of the operator) in the full space V.

(2) Another example of an unbounded operator is the one defined on $L^2(\mathbb{R})$ by

$$(Qf)(x) := x f(x) \tag{4.13}$$

which is very familiar from elementary wave mechanics. This is clearly unbounded (Exercise) and it cannot be defined on all of $L^2(\mathbb{R})$ since there are obviously some square-integrable functions that cease to be so when they are multiplied by "x".

(3) A well-known operator on l_2 is the *shift-operator* T defined on a sequence $(a_1, a_2, \ldots)$ as $T(a_1, a_2, \ldots) := (0, a_1, a_2, \ldots)$. To see that this is bounded, note that

$$\| T(a_1, a_2, \ldots) \|^2 = \| (0, a_1, a_2, \ldots) \|^2$$

$$= \sum_{i=1}^\infty |a_i|^2 = \| (a_1, a_2, \ldots) \|^2 .$$

Thus if $\mathbf{a} := (a_1, a_2, \ldots)$ denotes a general vector in l_2 we see that $\| T\mathbf{a} \| = \| \mathbf{a} \|$ which implies that $\| T \| = 1$.

It is a sad fact, but many operators in quantum theory are unbounded. For example, if A and B are any pair of operators that satisfy the Heisenberg commutation relations

$$[A, B] = i\mathbf{1}$$

then it can be shown that at least one of them must be unbounded. All this makes the rigorous formulation of quantum theory rather difficult.

2.5. UNITARY OPERATORS

We have already mentioned several times the importance of the general idea of a "morphism" as a structure preserving map between a pair of sets equipped with the same class of mathematical structure. In the case of vector spaces, such a map is called a "linear map" or, where the map is from a vector space into itself, a "linear operator". If the vector spaces are equipped with norms or scalar products, we might anticipate that the concept of a morphism could be usefully extended to include the requirement that the linear map/operator should also preserve the norm or scalar product. This is indeed the case and, for our purposes, particular significance is attached to a linear operator from a Hilbert space to itself that preserves the inner product structure. Such a so-called "unitary operator" play a fundamental role in the general mathematical structure of quantum theory and are best introduced simultaneously with the, equally important, idea of the "adjoint" of an operator.

Definitions.

(a) Let A be a bounded operator on a Hilbert space $\mathcal{H}$. Then the *adjoint* $A^\dagger$ of A is the unique operator on $\mathcal{H}$ satisfying

$$\langle \mathbf{v}, A^\dagger \mathbf{w} \rangle = \langle A\mathbf{v}, \mathbf{w} \rangle \tag{5.1}$$

for all vectors $\mathbf{v}$ and $\mathbf{w}$ in $\mathscr{H}$.

(b) The operator A is said to be *hermitian* if $A = A^\dagger$ or, equivalently,

$$\langle \mathbf{v}, A\mathbf{w} \rangle = \langle A\mathbf{v}, \mathbf{w} \rangle \quad \text{for all } \mathbf{v} \text{ and } \mathbf{w} \text{ in } \mathscr{H}. \tag{5.2}$$

(c) An invertible operator U is said to be *unitary* if it preserves the scalar product in the sense that

$$\langle U\mathbf{v}, U\mathbf{w} \rangle = \langle \mathbf{v}, \mathbf{w} \rangle \quad \text{for all } \mathbf{v} \text{ and } \mathbf{w} \text{ in } \mathscr{H}. \tag{5.3}$$

Note. (1) For all operators A and B,

$$(AB)^\dagger = B^\dagger A^\dagger. \tag{5.4}$$

(2) For a hermitian operator, the matrix elements with respect to an orthonormal basis, as defined in Eq. (4.9), satisfy

$$A_{ij} := \langle \mathbf{e}^i, A\mathbf{e}^j \rangle = \langle A\mathbf{e}^i, \mathbf{e}^j \rangle = \langle \mathbf{e}^j, A\mathbf{e}^i \rangle^* = A_{ji}^*,$$

i.e., they form a *hermitian* matrix.

(3) If U is a unitary operator then, for all $\mathbf{v}$ and $\mathbf{w}$ in $\mathscr{H}$,

$$\langle \mathbf{v}, \mathbf{w} \rangle = \langle U\mathbf{v}, U\mathbf{w} \rangle = \langle \mathbf{v}, U^\dagger U\mathbf{w} \rangle$$

and so,

$$U^\dagger U = 1 \tag{5.5}$$

which means in particular, since U has an inverse, this inverse is $U^{-1} = U^\dagger$ and that $UU^\dagger = 1$ also.

(4) The matrix elements of a unitary operator satisfy

$$\langle e^j, e^k \rangle = \langle Ue^j, Ue^k \rangle$$

$$= \sum_{i=1}^{\infty} \langle Ue^j, e^i \rangle \langle e^i, Ue^k \rangle \quad \text{from Eq. (3.27),} \quad (5.6)$$

$$= \sum_{i=1}^{\infty} \langle e^i, Ue^j \rangle^* \langle e^i, Ue^k \rangle$$

and so,

$$\delta^{jk} = \sum_{i=1}^{\infty} U_{ij}^* U_{ik}. \quad (5.7)$$

Note that if $\mathcal{H}$ has a finite dimension n then, the sums in Eqs. (5.6–7) are over an index ranging from $1, \ldots, n$, and in that case Eq. (5.7) asserts that the matrix representative of the unitary operator is a unitary $n \times n$ matrix. Thus, in terms of the concepts introduced around Eqs. (4.7–8), the matrix version of a unitary operator belongs to the $U(n)$ subgroup of the monoid $M(n, \mathbb{C})$.

(5) In general, the set of all unitary operators on a Hilbert space $\mathcal{H}$ forms a group that is a subset of the monoid of all *bounded* operators on $\mathcal{H}$. If $\mathcal{H} \cong \mathbb{C}^n$ then, as we have just remarked, this group is isomorphic to $U(n)$. If $\mathcal{H}$ is a real vector space, isomorphic to $\mathbb{R}^n$, then the group of unitary operators is isomorphic to the group $O(n, \mathbb{R})$. Note that, when $\mathcal{H}$ is real, it is more common to refer to a unitary operator as an "orthogonal" operator.

One of the most effective ways of studying hermitian or unitary operators is via their eigenvalues and eigenvectors.

Definition.

An *eigenvector* of a linear operator A is a vector **w** (non-null) in the vector space V such that

$$A\mathbf{w} = \mu\mathbf{w} \qquad (5.8)$$

for some complex number μ known as the *eigenvalue* of the operator associated with the eigenvector **w**.

Note. Care is needed if the eigenvalues belong to a continuous range. For then the concept of an eigenvector has to be revised somewhat to allow for the possibility that it will not be an element in the original vector space V but rather in a certain extension of it. For example, the differential operator $-i\,d/dx$ is a linear operator on the Hilbert space $L^2(\mathbb{R})$ with a familiar set of eigenvalues and eigenvectors

$$-i\frac{d}{dx}\,e^{ikx} = ke^{ikx} \qquad \text{for any } k \text{ in } \mathbb{R}.$$

But these "eigenvectors" $\mathbf{w}(x) := e^{ikx}$ do not lie in the space $L^2(\mathbb{R})$ since they are not square integrable.

The eigenvectors and eigenvalues of unitary and hermitian operators have some very special properties that are of great importance for the general theory of such operators and for their application in quantum theory. These are usually proved for hermitian operators so, by way of a change, we will give the theorem for the unitary case.

Theorem.

(a) The eigenvalues of a unitary operator U are of the form $e^{i\theta}$, i.e., they are complex numbers of modulus one.

(b) Two eigenvectors of U corresponding to two different eigenvalues are orthogonal to each other.

Proof.

(a) Let $U\mathbf{w} = \mu\mathbf{w}$. Then $\langle U\mathbf{w}, U\mathbf{w} \rangle = \langle \mu\mathbf{w}, \mu\mathbf{w} \rangle = \mu^*\mu \langle \mathbf{w}, \mathbf{w} \rangle = \langle \mathbf{w}, \mathbf{w} \rangle$.

Hence, since the case $\mathbf{w} = 0$ is excluded from the definition of an eigenvector, it follows that $\mu^*\mu = 1$.

(b) Now let μ_1 and μ_2 be two different eigenvalues of U with eigenvectors $\mathbf{w}_1$ and $\mathbf{w}_2$, respectively. Then

$$\langle \mathbf{w}_1, \mathbf{w}_2 \rangle = \langle U\mathbf{w}_1, U\mathbf{w}_2 \rangle = \langle \mu_1\mathbf{w}_1, \mu_2\mathbf{w}_2 \rangle = \mu_1^*\mu_2 \langle \mathbf{w}_1, \mathbf{w}_2 \rangle$$

which implies that $\langle \mathbf{w}_1, \mathbf{w}_2 \rangle = 0$ since $\mu_1^*\mu_2 = \mu_2/\mu_1$, which does not equal to 1 since $\mu_1 \neq \mu_2$.

$$\text{QED.}$$

Exercise. Show that a similar theorem holds for hermitian operators except that the eigenvalues are now real numbers.

There are many different ways in which group theoretic methods can be applied to the study of quantum theory. One of the longest standing of these requires the introduction of the concept of the "degeneracy" of an eigenvector (all eigenvalues are assumed to be discrete in this section).

Definition.

An eigenvalue μ of a linear operator A is said to be *d-fold* ($d \leq \infty$) *degenerate* if there exists a set of d linearly independent eigenvectors $\{\mathbf{w}^1, \mathbf{w}^2, \ldots, \mathbf{w}^d\}$ of A, all of which have the same eigenvalue μ. (d is assumed to be the largest such value.)

Note. (1) It is clear that $A(a_1\mathbf{w}^1 + a_2\mathbf{w}^2 + \ldots + a_n\mathbf{w}^n) =$

$\mu(a_1\mathbf{w}^1 + a_2\mathbf{w}^2 + \ldots + a_n\mathbf{w}^n)$, $n < \infty$ for any set of complex numbers $a_1, a_2, \ldots, a_n$. Hence, the set of *all* eigenvectors with the eigenvalue μ forms a d-dimensional linear subspace of V (which is closed if A is continuous).

(2) Let us label the *different* eigenvalues of A as $\mu_1, \mu_2, \ldots$ with the corresponding degeneracy numbers being d_1, $d_2, \ldots$. We will label the basis of eigenvectors with eigenvalue μ_m as $\{\mathbf{w}_m^1, \mathbf{w}_m^2, \ldots, \mathbf{w}_m^{d_m}\}$ and we can always choose these to be orthonormal so that

$$\langle \mathbf{w}_m^i, \mathbf{w}_m^j \rangle = \delta^{ij} \qquad i, j = 1, \ldots, d_m.$$

Then, from the previous theorem, if A is unitary (or hermitian) we have

$$\langle \mathbf{w}_m^i, \mathbf{w}_n^j \rangle = \delta^{ij} \delta_{mn} \qquad \text{for } i, j = 1, \ldots, d_n \quad (5.9)$$

and where m and n range from $1, \ldots, N$, where N is the number (possibly infinite) of *different* eigenvalues of A.

(3) A result of fundamental importance is the, so-called, *spectral theorem* which asserts that if A is unitary or hermitian, then the complete set of vectors $\{\mathbf{w}_m^i, i = 1, \ldots, d_m, m = 1, \ldots, N\}$ is an orthonormal basis set for the vector space. Thus any vector $\mathbf{v}$ can be expanded as:

$$\mathbf{v} = \sum_{m=1}^{N} \sum_{i=1}^{d_m} \langle \mathbf{w}_m^i, \mathbf{v} \rangle \, \mathbf{w}_m^i. \qquad (5.10)$$

Clearly, Dim $(\mathcal{H}) = \Sigma_{m=1}^{N} d_m$.

2.6. THE USE OF UNITARY OPERATORS IN QUANTUM THEORY

We shall now give a brief introduction to the way in which unitary operators tend to appear in quantum theory and the significance of group representations. We will start by stating the fundamental vector space approach to formulating a general mathematical framework in which to discuss quantum theory. This is contained in the following two rules.

Rule 1.

In a quantum theory, the states of the system (or, perhaps better, an ensemble of systems) are represented mathematically by vectors of unit length in a separable Hilbert space $\mathcal{H}$.

Rule 2.

An observable O is represented mathematically by a hermitian operator O acting on the Hilbert space of states $\mathcal{H}$.

If a measurement is made of this observable, the result obtained will necessarily be one of the eigenvalues of the associated operator $\hat{O}$. However, quantum theory is intrinsically incapable of saying, in general, *which* eigenvalue will occur in any given measurement. It can only predict the *probability* that a particular value will arise. This probability is given by the expression

$$\text{Prob}(O = \mu_m ; \mathbf{v}) = \sum_{i=1}^{d_m} |\langle \mathbf{e}_m^i, \mathbf{v} \rangle|^2 \qquad (6.1)$$

which represents the probability that a particular eigenvalue μ_m will be measured given that the state vector is $\mathbf{v}$. In Eq. (6.1), the vectors $\{\mathbf{e}_m^i, i = 1, \ldots, d_m, m = 1, \ldots, N\}$ are the eigenvectors of $\hat{O}$ corresponding to the eigenvalue μ_m which is d_m-fold degenerate. [cf. Eq. (5.10) and surrounding discussion.]

There is of course much that needs to be said about this pair of rules relating to the concepts of 'probability', 'measurement', 'observable' etc. However, as this is a course on group theory/vector space theory, rather than on quantum theory, I will restrict myself to commenting on some of the more mathematical implications of the formalism.

Comments.

(a) If the structure is to make any sense at all physically, it is necessary for the probability that *some* result is measured be equal to one. This involves summing the probabilities in Eq. (6.1) over all values of $m = 1, \ldots, N$ and, if we do, we find

$$\sum_{m=1}^{N} \text{Prob}(O = \mu_m ; \mathbf{v}) = \sum_{m=1}^{N} \sum_{i=1}^{d_m} \langle \mathbf{v}, \mathbf{e}_m^i \rangle \langle \mathbf{e}_m^i, \mathbf{v} \rangle$$
$$= \langle \mathbf{v}, \mathbf{v} \rangle,$$

where the last step follows from Parseval's formula in Eq. (3.26).

Since the state vectors were specified to have unit length/norm we see that the probabilities do indeed sum up to one. This demonstrates rather strikingly the importance of the expansion theorems of Sec. 2.3 in the formulation of a consistent mathematical framework for quantum theory.

(b) The average value of the results obtained from the measurements of the observable O is predicted by Eq. (6.1) to be

$$\langle O \rangle_{\mathbf{v}} = \sum_{m=1}^{N} \mu_m \text{Prob}(O = \mu_m ; \mathbf{v})$$

$$= \sum_{m=1}^{N} \sum_{i=1}^{d_m} \mu_m \langle \mathbf{v}, \mathbf{e}_m^i \rangle \langle \mathbf{e}_m^i, \mathbf{v} \rangle = \sum_{m=1}^{N} \sum_{i=1}^{d_m} \langle \mathbf{v}, \mathbf{e}_m^i \rangle \langle \hat{O} \mathbf{e}_m^i, \mathbf{v} \rangle$$

$$= \langle \mathbf{v}, \hat{O} \mathbf{v} \rangle \quad \text{from Eq. (3.27).}$$

This result

$$\langle O \rangle_v = \langle v, \hat{O}v \rangle \tag{6.2}$$

is one of the fundamental results of quantum theory.

(c) This procedure can in fact be reversed. Thus it is common to *start* with Eq. (6.2) as the basic interpretive rule of quantum physics. One can then prove as a mathematical consequence of the properties of hermitian operators that

   (i) The result of any measurement of O *must* be one of the eigenvalues of $\hat{O}$.

   (ii) The probability that a particular value μ_m will be obtained is given precisely by Eq. (6.1).

This alternative way of formulating Rule 2 has the advantage that it is readily extendible to a wider class of state than that described so far. Viz the "mixed" states which describe the situation in which it is not known with certainty which is the actual state vector of the system but only that it lies in a particular set of vectors with a certain probability (in the "ordinary" sense of the word).

Now we want to consider the way in which a unitary operator can reflect an arbitrariness in the specific association of states with vectors, and observables with hermitian operators. So, let U be any unitary operator on the Hilbert space $\mathcal{H}$. If μ is an eigenvalue of $\hat{O}$ with eigenvector w then,

$$U\hat{O}U^{-1}(Uw) = U(\hat{O}w) = U(\mu w) = \mu(Uw) \tag{6.3}$$

from which it follows that the hermitian operator $U\hat{O}U^{-1}$ $[= U\hat{O}U^{\dagger}]$ has exactly the same eigenvalues as the original operator $\hat{O}$. Furthermore, since U is a unitary operator we have

$$|\langle Uv, Uw \rangle|^2 = |\langle v, w \rangle|^2 \qquad \text{for all } v \text{ and } w \text{ in } \mathcal{H},$$

and,

$$\langle U\mathbf{v}, (U\hat{O}U^{-1})\, U\mathbf{w}\rangle = \langle U\mathbf{v}, U\hat{O}\mathbf{w}\rangle$$
$$= \langle \mathbf{v}, \hat{O}\mathbf{w}\rangle .$$

Hence, we obtain the result that, if a specific association has been made of

$$\begin{cases} \text{Physical state} & \longrightarrow \text{state vector } \mathbf{v} \text{ in } \mathcal{H} \\ \text{Physical observable } O \longrightarrow \text{hermitian operator } \hat{O} \text{ acting on } \mathcal{H} \end{cases}$$

then the same physical predictions will be obtained if we use instead the association

$$\begin{cases} \text{Physical state} & \longrightarrow \text{state vector } U\mathbf{v} \text{ in } \mathcal{H} \\ \text{Physical observable } O \longrightarrow \text{hermitian operator } U\hat{O}U^{-1} . \end{cases} \quad (6.4)$$

Thus we arrive at the important conclusion that the association of "physical entities" with their "mathematical representatives" in quantum theory is only defined up to arbitrary transformations of the type in Eq. (6.4).

Cases arise in quantum theory in which, for clear physical reasons, we expect to get different [but equivalent in the sense of Eq. (6.4)] mathematical representations of the same underlying physical states and observables.

Example.

Let two observers Fred and Fred' study the same quantum system using the same basic rules for assigning mathematical objects to physical quantities but let us assume that they use two different reference frames in physical 3-space. Then it seems reasonable to expect that a given physical state of the system will be described by them using different state vector (but in the same Hilbert space) since the system looks different in certain ways when viewed from their particular perspectives.

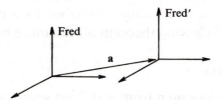

For example, in the wave mechanical description of a single particle, if the particle is prepared to be in a state described by the wavefunction $\psi(x)$ as viewed by Fred, then Fred' will wish to assign it the wavefunction $\psi'(x) := \psi(x + a)$. (**Note.** Both observers are using the symbol "x" to denote the vector position of the particle as measured from the origin of their own particular reference frame; note that these are two *different* physical observables.)

No physical results can depend on the choice of observer (at least, not in the simple Newtonian view of space that is being employed here) and hence if,

(a) the state vectors assigned by Fred and Fred' are **v** and **v'**, respectively;

(b) both observers measure the same physical observable O in which the eigenstates corresponding to a particular eigenvalue μ are assigned by them to be the eigenvectors **e** and **e'** of the operators $\hat{O}$ and $\hat{O}'$ which they respectively use,

then

$$|\langle \mathbf{e}, \mathbf{v} \rangle| = |\langle \mathbf{e'}, \mathbf{v'} \rangle| \qquad (6.5)$$

in order that both observers predict the same probabilities of getting the various common results for the value of the observable O.

Of course, one way of ensuing that Eq. (6.5) is satisfied is for **v** and **v'** [and **e** and **e'**] to be related as in Eq. (6.4). However, there seems no *a priori* reason why the map $\mathbf{v} \rightsquigarrow \mathbf{v'}$ satisfying Eq. (6.5) should even be *linear*, let alone described by a unitary operator. This makes the following theorem all the more remarkable.

Theorem (Wigner).

Let $\mathbf{v} \rightsquigarrow \mathbf{v}'$ be any map from a Hilbert space $\mathcal{H}$ to itself that is

(a) invertible; and

(b) such as to satisfy $|\langle \mathbf{v}, \mathbf{w} \rangle| = |\langle \mathbf{v}', \mathbf{w}' \rangle|$ (6.6)
for all $\mathbf{v}$ and $\mathbf{w}$ in $\mathcal{H}$.

Then there exists either a unitary operator or an anti-unitary operator $U : \mathcal{H} \to \mathcal{H}$ such that $\mathbf{v}' = U\mathbf{v}$ for all vectors $\mathbf{v}$ in $\mathcal{H}$.

Note. This important theorem is much too lengthy to prove here. But it is worth noting that

(a) An *anti-unitary* operator A is a map from $\mathcal{H}$ into itself that is
   (i) anti-linear in the sense that

$$A(\mu_1 \mathbf{v}_1 + \mu_2 \mathbf{v}_2) = \mu_1^* A\mathbf{v}_1 + \mu_2^* A\mathbf{v}_2 \qquad (6.7)$$

for all complex numbers μ_1, μ_2 and all vectors $\mathbf{v}_1$ and $\mathbf{v}_2$.
   (ii) such that, for all vectors $\mathbf{v}$ and $\mathbf{w}$ in $\mathcal{H}$,

$$\langle A\mathbf{v}, A\mathbf{w} \rangle = \langle \mathbf{w}, \mathbf{v} \rangle = \langle \mathbf{v}, \mathbf{w} \rangle^* . \qquad (6.8)$$

(b) The operator U is only defined up to a phase factor, i.e., $e^{i\theta} U$ works just as well as U.

Example.

Let G be any group of transformations of the set of reference frames in physical 3-space as discussed in Sec. 1.4. Thus this could be the group $\mathbb{R}^3$ of translations, the group $O(3, \mathbb{R})$ of rotations, or the full Euclidean group $\mathbb{R}^3 \circledS O(3, \mathbb{R})$ with the group law of Eq. (1.4.14).

Then to each g in G we get a unitary (or anti-unitary) operator U_g such that the state vector $\mathbf{v}'$ ascribed by the transformed observer (Fred') is related to the state vector $\mathbf{v}$ ascribed by the untransformed observer, by $\mathbf{v}' = U_g \mathbf{v}$. If we consider two successive transformations of reference frame corresponding to the group elements g_1 and g_2, respectively, then

$$\mathbf{v}' = U_{g_1} \mathbf{v} \quad \text{and} \quad \mathbf{v}'' = U_{g_2} \mathbf{v}' = U_{g_2} U_{g_1} \mathbf{v}.$$

On the other hand, we should get the same physical results if we transform at once to Fred'' with the group element $g_2 g_1$.
Thus

$$\mathbf{v}'' = \text{phase-factor} \cdot U_{g_2 g_1} \mathbf{v}$$

and hence we conclude that

$$\boxed{U_{g_2} U_{g_1} = \exp i\, \omega(g_2, g_1)\, U_{g_2 g_1}} \qquad (6.9)$$

in which the phase factor $\exp i\, \omega(g_2, g_1)$ may depend on both g_1 and g_2. A set of unitary operators $\{U_g, g \text{ in } G\}$ satisfying Eq. (6.9) is said to form a *projective representation* of the group G.

Note. (1) We must have $U_e = \mathbf{1}$, which is, of course, a unitary operator. This means that all group elements g "near" to the unit element e (more precisely, that can be connected to e by a continuous path in the Lie group G) must also be unitary. In principle, it is possible for group elements "disconnected" from e to be represented by an anti-unitary operator and for some purposes, such as the transformation of the time axis $t \rightsquigarrow -t$, this is in fact necessary. However, we will ignore that possibility here and from now on we will assume that U_g is a unitary operator.

(2) The operator U_g is only defined up to a phase factor $e^{i\theta(g)}$. If we replace the operators satisfying (6.9) with the redefined operators $e^{i\theta(g)} U_g$, the relations (6.9) become

$$U_{g_2} U_{g_1} = \exp i \left[\omega(g_2, g_1) - \theta(g_2 g_1) + \theta(g_2) + \theta(g_1) \right] U_{g_2 g_1}.$$
(6.10)

For many groups G, it is possible to choose the arbitrary phases $\theta(g)$ so that the total phase-factor in Eq. (6.10) vanishes

$$\omega(g_2, g_1) - \theta(g_2 g_1) + \theta(g_2) + \theta(g_1) = 0$$

and Eq. (6.10) becomes

$$\boxed{U_{g_2} U_{g_1} = U_{g_2 g_1}.}$$
(6.11)

Thus the map $g \rightsquigarrow U_g$ gives us a homomorphism from the group G into the group of all unitary operators on the Hilbert space $\mathcal{H}$. Such a map is called a *unitary representation* of the group G on the Hilbert space. This is a key idea in group theory and is the subject of the third and final major section of this lecture course.

3. GROUP REPRESENTATIONS

3.1. BASIC DEFINITIONS

In Sec. 1.4, we introduced the general idea of a group G acting on a set X via a homomorphism from G into the group $\text{Perm}(X)$ of bijections of X.

A very special, but extremely important, example is when X is a vector space and the homomorphism maps elements of G into the group $\text{Aut}(V)$ of invertible linear transformations of this space. We have already seen in the previous section how such a situation can arise naturally in quantum theory [eg. Eq. (2.6.11)] and now we must develop this idea properly.

Definitions.

(a) A *linear representation* of a group G on a vector space V is a homomorphism $T : G \to \text{Aut}(V)$. Thus for each g in G, $T(g)$ is an invertible operator and, for all g_1 and g_2 in G,

$$T(g_2 g_1) = T(g_2)\, T(g_1) . \qquad (1.1)$$

(b) A *unitary representation* of a group G on a Hilbert space $\mathcal{H}$ is a homomorphism U from G into the group of unitary operators on $\mathcal{H}$. Thus for each g in G, $U(g)$ is a unitary operator and, for all g_1 and g_2 in G,

$$U(g_2 g_1) = U(g_2)\, U(g_1)\,. \tag{1.2}$$

[In Eq. (2.6.11), we wrote U_g rather than $U(g)$. These two symbols are interchangeable.]

(c) As discussed in Sec. 1.5, it is useful to introduce the idea of the kernel of T [or U] as the normal subgroup of G that is mapped into the unit element in $\text{Aut}(V)$ and is hence 'not represented' by this homomorphism. In particular, the representation is said to be *faithful* if its kernel is trivial, i.e., $\ker T = \{e\}$.

Now we must consider the important question of when two different linear representations of G can be regarded as being equivalent. If the two representations are T_1 and T_2 defined on vector spaces V_1 and V_2, respectively, then one necessary condition for equivalence is that V_1 and V_2 should be isomorphic. However, this is not sufficient as we also require that, in an appropriate sense, the group G is represented by T_1 and T_2 in "the same way". What we are searching for is the correct definition of a morphism between two linear representations; an 'isomorphism' will then be defined as usual as a bijective morphism. The morphism should clearly be a morphism between the vector spaces V_1 and V_2 (i.e., a linear map) which preserves the group actions. Such a map is called an 'intertwining operator' and is defined as follows.

Definitions.

(a) If T_1 and T_2 are linear representations of a group G on vector spaces V_1 and V_2, respectively, then an *intertwining operator* between these representations is a linear map $A : V_1 \to V_2$ such that the following diagram is commutative

$$V_1 \xrightarrow{\quad A \quad} V_2$$

$$\downarrow T_1(g) \qquad\qquad \downarrow T_2(g)$$

$$V_1 \xrightarrow{\quad A \quad} V_2$$

i.e., $T_2(g)A = AT_1(g)$ for all g in G. (1.3)

(b) If the linear map A is an isomorphism between the vector spaces V_1 and V_2 then the two representations are said to be *equivalent* representations.

If U_1 and U_2 are both unitary representations on Hilbert spaces $\mathscr{H}_1$ and $\mathscr{H}_2$, respectively, and if the intertwining isomorphism A also preserves the inner products

$$\langle A\mathbf{v}, A\mathbf{w} \rangle_{\mathscr{H}_2} = \langle \mathbf{v}, \mathbf{w} \rangle_{\mathscr{H}_1} \qquad \text{for all } \mathbf{v} \text{ and } \mathbf{w} \text{ in } \mathscr{H}_1 \quad (1.4)$$

then the two representations are said to be *unitarily equivalent*. One of the most fundamental problems in group theory is to classify unitary representations of a group up to unitary equivalence.

Note. (1) If T is a linear representation of a group G on a vector space V and if A is any invertible operator on V [i.e., A belongs to Aut(V)] then $T'(g) := A^{-1} T(g) A$ is also a representation of G and it is clearly equivalent to T

$$V \xrightarrow{\quad A \quad} V$$

$$\downarrow T'(g) \qquad\qquad \downarrow T(g) \qquad\qquad (1.5)$$

$$V \xrightarrow{\quad A \quad} V$$

A is sometimes said to be a *similarity transformation*.

(2) If T is a linear representation of G on a finite-dimensional vector space $\mathbb{C}^n$ then T is a homomorphism from G into the Lie group of matrices $GL(n, \mathbb{C})$. This is known as a *matrix representation* of G. More generally, if T is a linear representation on *any* finite-dimensional vector space V then, as explained in Sec. 2.1, any basis set $\{e^1, e^2, \ldots, e^n\}$ of V gives rise to an isomorphism $i: V \to \mathbb{C}^n$ in which $i(\Sigma v_i e^i) := (v_1, v_2, \ldots, v_n)$. As discussed in Sec. 2.4, a linear operator such as $T(g)$ on V generates a matrix version on $\mathbb{C}^n$ which we will denote by $M(g)$ and which is defined as $M(g) := iT(g)i^{-1}$. Clearly, $g \rightsquigarrow M(g)$ is a matrix representation of G in $GL(n, \mathbb{C})$ and it is equivalent to the original representation T by construction [cf. Eq. (2.4.8)]

$$
\begin{array}{ccc}
V & \xrightarrow{\quad i \quad} & \mathbb{C}^n \\
\downarrow{\scriptstyle T(g)} & & \downarrow{\scriptstyle M(g)} \\
V & \xrightarrow{\quad i \quad} & \mathbb{C}^n
\end{array}
\qquad (1.6)
$$

Exercise. (1) Show that the matrix versions $M(g)$ and $M'(g)$ of $T(g)$ corresponding to two different choices of basis for V are related by a similarity transformation.

(2) If U is a finite-dimensional unitary representation of G then show that it is equivalent to a matrix representation of G in the Lie group $U(n)$, where $n = \dim(V)$.

(3) If $T: G \to \text{Aut}(V)$ is a representation of G then, since $\text{Ker } T$ is a normal subgroup of G, $G/\text{Ker } T$ is itself a group and, from the discussion in Sec. 1.5, it follows that T induces a representation of $G/\text{Ker } T$ in $\text{Aut}(V)$ and that this representation is faithful.

(4) Many groups have no faithful, finite-dimensional uni-

tary representations at all. Examples are the linear groups $GL(n, \mathbb{C})$ and $GL(n, \mathbb{R})$, the euclidean group $\mathbb{R}^n \circledS O(n, \mathbb{R})$ and the Weyl-Heisenberg group with the group law in Eq. (1.3.22). The last two examples are one way of understanding why infinite-dimensional Hilbert spaces are an intrinsic feature of wave mechanics: the euclidean group $\mathbb{R}^3 \circledS O(3, \mathbb{R})$ describes the relation between different reference frames and is represented on the quantum state space according to the discussion in Sec. 2.6, and a unitary representation of the Weyl-Heisenberg group is equivalent to a representation by hermitian operators of the fundamental canonical commutation relations in Eq. (1.3.23).

Examples.

(1) A one-dimensional representation of the group $\mathbb{Z}_2 = \{e, a\}$ on the vector space $\mathbb{C}$, is given by

$$U(e) := 1 , \quad U(a) := -1 . \tag{1.7}$$

A two-dimensional representation on $\mathbb{C}^2$ is

$$U(e) := \begin{pmatrix} 1 & 0 \\ 0 & 1 \end{pmatrix}, \quad U(a) := \begin{pmatrix} 0 & 1 \\ 1 & 0 \end{pmatrix}. \tag{1.8}$$

Note that both these matrix representations are unitary, and they are both faithful.

(2) A representation of the general cyclic group $\mathbb{Z}_n = \{e, a, a^2, \ldots, a^{n-1}\}$ with $a^n = e$, can be constructed with the aid of any square matrix M, of any dimension, that satisfies the equation $M^n = 1$: we simply define $T(a^q) := M^q$ and $T(e) := 1$. For example, $\mathbb{Z}_4 = \{e, a, a^2, a^3\}$, and the 3×3 matrix

$$M := \begin{pmatrix} 1 & 0 & 0 \\ 0 & 0 & -1 \\ 0 & 1 & 0 \end{pmatrix} \text{ satisfies } M^4 = \begin{pmatrix} 1 & 0 & 0 \\ 0 & 1 & 0 \\ 0 & 0 & 1 \end{pmatrix} \quad (1.9)$$

and hence we get a 3-dimensional matrix representation of $\mathbb{Z}_4$ by

$$T(e) := \begin{pmatrix} 1 & 0 & 0 \\ 0 & 1 & 0 \\ 0 & 0 & 1 \end{pmatrix} \qquad T(a) := \begin{pmatrix} 1 & 0 & 0 \\ 0 & 0 & -1 \\ 0 & 1 & 0 \end{pmatrix}$$

$$\qquad\qquad\qquad\qquad\qquad\qquad\qquad\qquad\qquad (1.10)$$

$$T(a^2) := \begin{pmatrix} 1 & 0 & 0 \\ 0 & -1 & 0 \\ 0 & 0 & -1 \end{pmatrix} \qquad T(a^3) := \begin{pmatrix} 1 & 0 & 0 \\ 0 & 0 & 1 \\ 0 & -1 & 0 \end{pmatrix}$$

Clearly, Ker $T = \{e\}$, and hence the representation is faithful.

(3) Another example of a representation of $\mathbb{Z}_4$ can be obtained with the aid of the matrix $M = \begin{pmatrix} 0 & 1 \\ 1 & 0 \end{pmatrix}$ which also satisfies the equation $M^4 = \mathbf{1}$. This gives the 2×2 matrix representation on $\mathbb{C}^2$

$$T(e) := \begin{pmatrix} 1 & 0 \\ 0 & 1 \end{pmatrix} \qquad T(a) := \begin{pmatrix} 0 & 1 \\ 1 & 0 \end{pmatrix}$$

$$\qquad\qquad\qquad\qquad\qquad\qquad\qquad\qquad\qquad (1.11)$$

$$T(a^2) := \begin{pmatrix} 1 & 0 \\ 0 & 1 \end{pmatrix} \qquad T(a^3) := \begin{pmatrix} 0 & 1 \\ 1 & 0 \end{pmatrix}.$$

Note that Ker $T = \{e, a^2\}$ and hence this representation is not faithful. However, it does give us a faithful representation of the quotient group $\mathbb{Z}_4/\mathbb{Z}_2 \cong \mathbb{Z}_2$.

3.2. SYMMETRY TRANSFORMATIONS IN QUANTUM THEORY

As time passes, the vector representing the state of a quantum system will change, giving rise to a curve of vectors $t \rightsquigarrow \psi_t$ in the Hilbert space $\mathcal{H}$. The fundamental dynamical law of quantum theory is that this time evolution is described by the first-order differential equation

$$i\hbar \frac{d}{dt} \psi_t = H \psi_t, \qquad (2.1)$$

where the linear hermitian operator H — the *Hamiltonian* of the system — represents the energy observable. Any transformation of the system that leaves the Hamiltonian operator (and hence the dynamical evolution) alone is of great significance and motivates the following definition.

Definition.

Let $g \rightsquigarrow U_g$ be a faithful unitary representation of a group G on the Hilbert space of the system with Hamiltonian operator H. Then the group G is said to be a *symmetry* of the system if, for all g in G,

$$U_g H U_g^{-1} = H. \qquad (2.2)$$

Comments.

(a) From the discussion in Sec. 2.6, we know that one possible way in which such a family of unitary operators can arise is when we study the description of the quantum states/observables from the perspectives of two observers who use different reference frames that are related by the elements of some group of transformations G. If Eq. (2.2) is satisfied, it means that, in some sense, the quantum system (or at least its time development) 'looks the same' when viewed from all the reference frames that are related to each other in this way.

(b) We know from the discussion surrounding Eq. (2.6.3) that if **w** is an eigenvector of H with eigenvalue E then, for all g in G, U_g**w** will be an eigenvector of $U_g H U_g^{-1}$ with the same eigenvalue. However, if G is a symmetry of the system, then $U_g H U_g^{-1} = H$ and hence U_g**w** is actually an eigenvector of H with the same eigenvalue as the original eigenvector **w**. The important conclusion is that:

> For each E, the operators U_g map the subspace of degenerate eigenvectors of H into itself. Thus this space carries, by itself, a unitary representation of G. This opens up the general possibility of describing the degeneracy of the eigenvalues of any Hamiltonian in terms of the group representations of its symmetry groups.

(c) In relativistic quantum theory, it is more appropriate to discuss symmetries in terms of the behavior of the relativistically invariant entity $P_0^2 - \mathbf{P} \cdot \mathbf{P}$ where (in units of $c = 1$), $(P_0, \mathbf{P})$ are the four components of the energy-momentum 4-vector. In this case, a symmetry group is a group G with a unitary representation $g \rightsquigarrow U_g$ such that, for all g in G,

$$U_g (P_0^2 - \mathbf{P} \cdot \mathbf{P}) U_g^{-1} = P_0^2 - \mathbf{P} \cdot \mathbf{P}. \qquad (2.3)$$

The discrete eigenvalues of $P_0^2 - \mathbf{P} \cdot \mathbf{P}$ are just the possible (rest mass)^2 of the particles in the theory and, as in the case (b) above, the degeneracy subspaces will be mapped into each other under the action of G. This leads to the idea of classifying 'multiplets of particles' (i.e., sets of particles with the same mass) in terms of the representation theory of some symmetry group. This will not be a transformation group of spacetime reference frames but rather an 'internal symmetry group' which arises from the posited detailed structure of the system. Examples of such groups are the $SU(2)$ and $SU(3)$ groups used in strong interaction physics.

Examples.

(1) As an example of the idea of symmetry in non-relativistic quantum theory, let us return to the example in Sec. 2.6 of a pair of observers Fred and Fred' who inhabit reference frames related by a translation vector **a**.

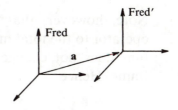

This gives rise to a unitary representation $\mathbf{a} \rightsquigarrow U_\mathbf{a}$ of the translation group of reference frames and, as we have discussed in Sec. 2.6, if Fred reckons the wavefunction is $\psi(\mathbf{x})$ then Fred' will wish to use the wavefunction $\psi'(\mathbf{x}) := \psi(\mathbf{x} + \mathbf{a})$. Thus we have

$$[U_\mathbf{a}\psi](\mathbf{x}) = \psi(\mathbf{x} + \mathbf{a}) \tag{2.4}$$

which is indeed a unitary representation of the translation group $\mathbb{R}^3$ on the Hilbert space $L^2(\mathbb{R}^3)$.

Now, let **X** denotes the physical observable that specifies the vector position of the particle with respect to the origin of Fred's reference frame. Then, using the usual quantum rules, Fred will represent this observable with the operator $\hat{\mathbf{x}}$ where

$$[\hat{x}_i\psi](\mathbf{x}) := x_i\psi(\mathbf{x}) \quad \text{for } i = 1, \ldots, 3. \tag{2.5}$$

On the other hand, the *same* physical observable will be represented by Fred' (using the same rules as Fred) by the operator

$$\hat{x}' := \hat{x} + a, \quad \text{i.e.,} \quad [\hat{x}'_i \psi](x) := (x_i + a_i) \psi(x) \quad (2.6)$$

since he uses "**x**" to denote the vector position from the origin of **his** reference frame. Thus we have

$$U_a \hat{x} U_a^{-1} = \hat{x}' = \hat{x} + a. \quad (2.7)$$

Note, however, that both observers will assign the *same* operator to the total momentum of the particle since momentum does not change under the translation of a reference frame. Hence,

$$U_a \hat{\mathbf{p}} U_a^{-1} = \hat{\mathbf{p}}. \quad (2.8)$$

Now, we can consider whether or not translations are a symmetry of various types of Hamiltonian. The simplest case is when the particle is free so that

$$H = \frac{\mathbf{p} \cdot \mathbf{p}}{2m}. \quad (2.9)$$

Then,

$$U_a(\hat{\mathbf{p}} \cdot \hat{\mathbf{p}}) U_a^{-1} = \sum_{i=1}^{3} U_a \hat{p}_i U_a^{-1} U_a \hat{p}_i U_a^{-1}$$

$$= \hat{\mathbf{p}} \cdot \hat{\mathbf{p}} \quad (2.10)$$

by virtue of Eq. (2.8). Thus, the translation group $\mathbb{R}^3$ *is* a symmetry group of the free particle.

On the other hand,

$$U_a(\hat{\mathbf{p}} \cdot \hat{\mathbf{p}}) \, U_a^{-1} = \sum_{i=1}^{3} U_a \, \hat{p}_i \, U_a^{-1} \, U_a \, \hat{p}_i \, U_a^{-1}$$

$$= (\hat{\mathbf{x}} + \mathbf{a}) \cdot (\hat{\mathbf{x}} + \mathbf{a}) \qquad (2.11)$$

by virtue of Eq. (2.7). Therefore, translations are *not*, for example, a symmetry group of a hydrogen atom whose nucleus is at the origin of Fred's reference frame and with the Hamiltonian

$$H = \frac{\mathbf{p} \cdot \mathbf{p}}{2m} - \frac{e^2}{(\mathbf{x} \cdot \mathbf{x})^{\frac{1}{2}}}. \qquad (2.12)$$

(2) However, if we look instead at the effects of rotating Fred's reference frame with a rotation matrix in $O(3, \mathbb{R})$, then the associated unitary representation on the wavefunction is [cf. Eq. (1.1)]

$$[U_R \psi] (\mathbf{x}) = \psi(\mathbf{R}^{-1}\mathbf{x})$$

and

$$U_R \, \hat{x}_i \, U_R^{-1} = \sum_{j=1}^{3} \hat{x}_j \, R_{ji}; \quad U_R \, \hat{p}_i \, U_R^{-1} = \sum_{j=1}^{3} \hat{p}_j \, R_{ji}$$

from which it follows that $\hat{\mathbf{x}} \cdot \hat{\mathbf{x}}$ and $\hat{\mathbf{p}} \cdot \hat{\mathbf{p}}$ are invariant, and hence so is the H-atom Hamiltonian in Eq. (2.12). This is deeply connected with the fact that the degenerate (in energy) subspaces carry angular momentum [the "*l*" and "*m*" quantum numbers effectively label different unitary representations of $O(3, \mathbb{R})$].

3.3. REDUCIBILITY OF REPRESENTATIONS

In discussing symmetry and eigenvalue degeneracy, we saw in the last section how it was possible for a representation of a group G on a vector space V to be such that there was some subspace of V that was mapped into itself by G, and which hence constituted a G-representation in its own right. This is a very important idea and we shall develop it in the present section. First, some formal definitions.

Definitions.

(a) Let T be a linear representation of a group G on a vector space V. A linear subspace W of V (closed if V is a Hilbert space) is *G-invariant* if, for all g in G and all $\mathbf{w}$ in W, $T(g)\mathbf{w}$ belongs to the subspace W, i.e., $T(g) : W \to W$ for all g in G.

The ensuing representation of G on W is called the *restriction* of T to W and is said to be a *subrepresentation* of T, which we shall write as T_W.

The quotient space V/W also carries a representation of G called the *quotient representation* $T_{V/W}$ and defined by

$$T_{V/W}(g)\,(\mathbf{v} + W) := T(g)\mathbf{v} + W. \qquad (3.1)$$

(b) A representation T of G on a vector space V is *irreducible* if the only G-invariant subspaces of V are $\{0\}$ and V itself. Otherwise, T is said to be *reducible*.

Example.

Let $M_1 : G \to GL(n_1, \mathbb{C})$ and $M_2 : G \to GL(n_2, \mathbb{C})$ be two matrix representations of G that are irreducible. Thus there are no proper invariant subspaces of the vector spaces $\mathbb{C}^{n_1}$ and $\mathbb{C}^{n_2}$ on which they are respectively defined. Now let us consider the set of all block diagonal $(n_1 + n_2) \times (n_1 + n_2)$ matrices of the form

$$M(g) := \left(\begin{array}{c|c} M_1(g) & 0 \\ \hline 0 & M_2(g) \end{array} \right) \tag{3.2}$$

which clearly satisfy $M(g) M(g') = M(gg')$ and hence constitute a matrix representation on the vector space of $(n_1 + n_2)$-tuples of complex numbers, i.e., on $\mathbb{C}^{(n_1 + n_2)}$. It is also clear that this new representation is *not* irreducible since the set of all vectors of the form $(a_1, a_2, \ldots, a_{n_1}, 0, 0, \ldots, 0)$ are mapped into themselves by the group action, as are the vectors of the form $(0, 0, \ldots, 0, a_{n_1+1}, a_{n_1+2}, \ldots, a_{n_1+n_2})$.

This last observation prompts an interesting question. Suppose that one of the representations, say M_1, was not irreducible but itself had proper invariant subspaces. Would it be possible with the aid of a similarity transformation to write M_1 in a reduced block diagonal form with the invariant subspaces clearly evident as in the case of Eq. (3.2)? One could go on doing this perhaps until the entire representation was written in a block diagonal form in which each block corresponded to a *irreducible* representation and hence could not be reduced further.

Groups for which every representation could be decomposed in this way would have a particularly tractable representation theory since the classification of arbitrary representations would reduce to the problem of classifying the irreducible representations, and then specifying how many times each one of these appeared in the 'block diagonal' decomposition of a general representation. This is the path we wish to pursue. But first, we must broaden the discussion to include general vector spaces in addition to the special ones of the form $\mathbb{C}^n$ that were used in the analysis of Eq. (3.2). The crucial step is to define the sense in which an arbitrary vector space can be 'split' into the 'sum' of subspaces as an analogue of the way in which the space $\mathbb{C}^{(n_1 + n_2)}$ splits into the 'sum' of $\mathbb{C}^{n_1}$ and $\mathbb{C}^{n_2}$ in the sense that every vector in $\mathbb{C}^{(n_1+n_2)}$ can be written uniquely as a sum of vectors in $\mathbb{C}^{n_1}$ and $\mathbb{C}^{n_2}$

$$(a_1, a_2, \ldots, a_{n_1}, a_{n_1+1}, \ldots, a_{n_1+n_2})$$

$$= (a_1, a_2, \ldots, a_{n_1}, 0, 0, \ldots, 0) \qquad (3.3)$$

$$+ (0, 0, \ldots, 0, a_{n_1+1}, a_{n_1+2}, \ldots, a_{n_1+n_2}).$$

The key idea is that of a "direct sum" of two vector spaces.

Definitions.

(a) The *direct sum* of two complex vector spaces V_1 and V_2 is the set of all pairs (v_1, v_2) with v_1 in V_1 and v_2 in V_2, with the vector space operations

$$(v_1, v_2) + (v_1', v_2') := (v_1 + v_1', v_2 + v_2') \qquad (3.4)$$

$$\mu(v_1, v_2) := (\mu v_1, \mu v_2). \qquad (3.5)$$

The direct sum of the two vector spaces is written as $V_1 \oplus V_2$.

(b) The direct sum of two Hilbert spaces $\mathcal{H}_1$ and $\mathcal{H}_2$ is defined as above and with a scalar product

$$\langle (v_1, v_2), (v_1', v_2') \rangle := \langle v_1, v_1' \rangle_{\mathcal{H}_1} + \langle v_2, v_2' \rangle_{\mathcal{H}_2}. \qquad (3.6)$$

(c) If T_1 and T_2 are two linear representations of G on vector spaces V_1 and V_2 respectively, then the *direct sum* representation is defined on the vector space $V_1 \oplus V_2$ by

$$[T_1 \oplus T_2](g)(v_1, v_2) := (T_1(g)v_1, T_2(g)v_2). \qquad (3.7)$$

Note. (1) The vector space $\mathbb{C}^n$ is isomorphic to the direct sum $\mathbb{C}^{n_1} \oplus \mathbb{C}^{n_2}$ for any pair of positive integers n_1 and n_2 such that $n = n_1 + n_2$. The isomorphism may be constructed immediately by studying Eq. (3.3)

$$(a_1, a_2, \ldots, a_{n_1+1}, a_{n_1+2}, \ldots, a_{n_1+n_2})$$
$$\rightsquigarrow ((a_1, a_2, \ldots, a_{n_1}), (a_{n_1+1}, \ldots, a_{n_1+n_2})) .$$

The block diagonal representation in Eq. (3.2) is then isomorphic to the direct sum $M_1 \oplus M_2$ of the two matrix representations M_1 and M_2.

(2) In general, $\dim(V_1 \oplus V_2) = \dim(V_1) + \dim(V_2)$.

(3) In $V_1 \oplus V_2$, any vector $(\mathbf{v}_1, \mathbf{v}_2)$ can be decomposed uniquely as

$$(\mathbf{v}_1, \mathbf{v}_2) = (\mathbf{v}_1, 0) + (0, \mathbf{v}_2) \tag{3.8}$$

which can be thought of as a sum of vectors in V_1 and V_2 if we regard V_1 and V_2 as subspaces of $V_1 \oplus V_2$ via the natural injections

$$i_1 : V_1 \rightarrow V_1 \oplus V_2 \qquad i_2 : V_2 \rightarrow V_1 \oplus V_2 \tag{3.9}$$
$$i_1(\mathbf{v}_1) := (\mathbf{v}_1, 0) \qquad i_2(\mathbf{v}_2) := (0, \mathbf{v}_2) .$$

Conversely, a vector space V is said to *decompose* into the direct sum of a pair of subspaces V_1 and V_2 if any $\mathbf{v}$ in V can be written uniquely in the form $\mathbf{v} = \mathbf{v}_1 + \mathbf{v}_2$ with $\mathbf{v}_1$ and $\mathbf{v}_2$ belonging to V_1 and V_2, respectively. This implies that

$$\text{(a) } V_1 \cap V_2 = \{0\}$$

$$\text{(b) } V \cong V_1 \oplus V_2.$$

(4) If W is any subspace of a vector space V, there will be many 'complementary' subspaces W' such that $V \cong W \oplus W'$. For example

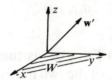

In $\mathbb{R}^3$, if W is the x–y plane, W' could be the one-dimensional subspace generated by any vector $\mathbf{w}'$ that does not lie in the x–y plane.

However, if $\mathscr{H}$ is a Hilbert space and W is a closed subspace, then a natural choice for W' is the orthogonal complement of W in $\mathscr{H}$

$$W_\perp := \{\mathbf{v} \text{ in } \mathscr{H} \text{ such that } \langle \mathbf{w}, \mathbf{v} \rangle = 0 \text{ for all } \mathbf{w} \text{ in } W\}.$$

This would correspond to picking $\mathbf{w}'$ to point along the z-axis in the diagram above. In the general case, the discussion preceding Eq. (3.29) showed that $\mathscr{H} \cong W \oplus W_\perp$.

As far as using direct sums of vector spaces to study group representations is concerned, the following remarks are in order.

Comments.

(a) If T_1 and T_2 are representations of G on V_1 and V_2 respectively, then the direct sum representation $T_1 \oplus T_2$ is certainly reducible since V_1 and V_2 are G-invariant subspaces [using the injections in Eq. (3.9)] of $V_1 \oplus V_2$.

(b) If W is a G-invariant subspace of a vector space V that carries a representation T of G, there may or may not be another G-invariant subspace W' of V such that $T \cong T_W \oplus T_{W'}$.

A representation is said to be *completely reducible* if, for any G-invariant subspace V, such a 'complementary' invariant subspace can be found. In this case, it is easy to show that $T_{V/W} \cong T_{W'}$ and of course T is uniquely specified by the two subrepresentations T_W and $T_{W'} \cong T_{V/W}$.

However, when T is *not* completely reducible, it is not necessarily the case that, given a G-invariant subspace W of V, the representation T can be 'recovered' from the representations T_W and $T_{V/W}$. In the case of a matrix representation, the matrices look like

$$T(g) = \left(\begin{array}{c|c} T_W(g) & B(g) \\ \hline 0 & T_{V/W}(g) \end{array} \right) \tag{3.10}$$

and there is no similarity transformation A such that $AT(g)A^{-1}$ is in block diagonal form for all g in G. This means that one needs to know something about the matrices $B(g)$ as well as the subrepresentation $T_W(g)$ and the quotient representation $T_{V/W}(g)$ in order to fully 'recover' the original representation $T(g)$.

The importance of knowing if a representation T is completely reducible is that it allows the possibility of expressing T as a direct sum of *irreducible* representations. Thus one chooses an invariant subspace W of V and decomposes the representation as $T \cong T_W \oplus T_{W'}$ where W' is some complementary invariant subspace of V. If T_W is not itself irreducible, then choose an invariant subspace Y of W (and similarly for W'), and so on. With a bit of luck, this procedure will terminate at some stage and one will be left with a decomposition of T into a direct sum of irreducible representations.

Note. (1) For some groups this procedure does *not* terminate even though the representation is completely reducible. Of course, this 'pathology' can only occur when the original representation is infinite-dimensional. We shall see a specific example of this shortly.

(2) Even if the procedure does terminate, it is necessary to show that the decomposition into a direct sum of irreducible representations does not depend on the choices of invariant subspaces that are made during the execution of the process.

It is evidently of considerable importance to know what groups and what representations are completely reducible. The following rather simple theorem is of great interest from this point of view.

Theorem.

Every unitary representation $g \rightsquigarrow U(g)$ on a Hilbert space $\mathscr{H}$ of any group G, is completely reducible.

Proof.

Let W be any G-invariant subspace of $\mathscr{H}$. We know that, as a vector space, $\mathscr{H} \cong W \oplus W_{\perp}$. The theorem will be proved if we can show that $W_{\perp}$ is also G-invariant.

Let $\mathbf{v}$ belong to $W_{\perp}$. Then, for any $\mathbf{w}$ in W, we have, for all g in G,

$$\langle U(g)\mathbf{v}, \mathbf{w} \rangle = \langle U(g)\mathbf{v}, U(g) \, U^{-1}(g)\mathbf{w} \rangle = \langle \mathbf{v}, U^{-1}(g)\mathbf{w} \rangle \quad (3.11)$$

since $U(g)$ is a unitary operator. But $U^{-1}(g)\mathbf{w} = U(g^{-1})\mathbf{w}$ which belongs to W since it is G-invariant. Thus $\langle \mathbf{v}, U^{-1}(g)\mathbf{w} \rangle = 0$ and hence, from Eq. (3.11), it follows that $\langle U(g)\mathbf{v}, \mathbf{w} \rangle = 0$ for all $\mathbf{w}$ in W, which implies that $U(g)\mathbf{v}$ belongs to $W_{\perp}$. Thus $\mathbf{v}$ in $W_{\perp}$ implies $U(g)\mathbf{v}$ in $W_{\perp}$ for all g in G, i.e., $W_{\perp}$ is G-invariant.

QED.

Examples.

(1) The unitary representation of $O(3, \mathbb{R})$ on the Hilbert space $L^2(\mathbb{R}^3)$ given by Eq. (1.16) as $[U(R)f](\mathbf{x}) := f(\mathbf{R}^{-1}\mathbf{x})$ is comple-

tely reducible according to the theorem just proved. Thus it decomposes into a direct sum of irreducible representations of $O(3, \mathbb{R})$. All such representations of this Lie group are finite-dimensional with dimensions $2j + 1$ for $j = 0, 1, 2, \ldots$. In fact, j is nothing but the total angular momentum quantum number, and the group theoretic direct sum decomposition of $L^2(\mathbb{R}^3)$ reproduces the familiar decomposition of this space according to the values of the total angular momentum.

(2) The non-unitary representation of the additive group of the integers $\mathbb{Z}$ given by

$$T(n) := \begin{pmatrix} 1 & 1 \\ 0 & 1 \end{pmatrix}^n = \begin{pmatrix} 1 & n \\ 0 & 1 \end{pmatrix} \tag{3.12}$$

is reducible since the set of all vectors in $\mathbb{C}^2$ of the form $\begin{pmatrix} a \\ 0 \end{pmatrix}$ for some complex number a, is an invariant subspace of the group action. But the representation is *not* completely reducible since there is no matrix $\begin{pmatrix} a & b \\ c & d \end{pmatrix}$ which can be used to perform a similarity transform to put the representation in diagonal form, i.e., the equations

$$\begin{pmatrix} a & b \\ c & d \end{pmatrix} \begin{pmatrix} 1 & 1 \\ 0 & 1 \end{pmatrix} \begin{pmatrix} a & b \\ c & d \end{pmatrix}^{-1} = \begin{pmatrix} \alpha & 0 \\ 0 & \beta \end{pmatrix} \tag{3.13}$$

have no solutions for any choice of the complex numbers α and β. (Exercise)

(3) A unitary representation of the additive group $\mathbb{R}$ of the real numbers can be defined on $L^2(\mathbb{R})$ by

$$[U(t)f](x) := e^{itx} f(x). \tag{3.14}$$

Since U is unitary, this representation must be completely reducible. Nevertheless, there are *no* irreducible subspaces at all, and this affords an illustration of the 'pathology' mentioned earlier in which the sequential decomposition into invariant subspaces never terminates. In this particular case, the direct sum decomposition needs to be replaced by a 'direct integral' of irreducible representations. This is a well-defined and important concept but it would take much too long to develop it any further here.

In the next theorem, and for most of the results in the rest of the lecture course, a key step in the proof will involve summing over all elements in a finite group. This technique will not work for a countably infinite group as there is no guarantee that the sum will converge (neither is it always clear in what sense convergence would be understood). It is very important to know however, that for compact Lie groups [ones whose group spaces are bounded; like SU(2) which is a 3-sphere] the technique *does* work but with the sum replaced with a certain integral over the bounded group space.

Theorem.

Let T be a representation of a finite group G on a finite dimensional vector space V. Then there exists an inner product on V such that T is a unitary representation.

Proof.

Let $\langle \, , \, \rangle$ be any scalar product on V [it certainly has one; eg. use an isomorphism with $\mathbb{C}^n$ ($n = \dim(V)$) to employ the standard scalar product on $\mathbb{C}^n$]. Define

$$\langle \mathbf{v}, \mathbf{w} \rangle^{\sim} := \frac{1}{|G|} \sum_{g' \text{ in } G} \langle T(g')\mathbf{v}, T(g')\mathbf{w} \rangle . \qquad (3.15)$$

Then, $\langle \, , \, \rangle^\sim$ is also a scalar product on V (Exercise).
Furthermore:

$$\langle T(g)\mathbf{v}, T(g)\mathbf{w}\rangle^\sim = \frac{1}{|G|} \sum_{g' \text{ in } G} \langle T(g')\, T(g)\mathbf{v}, T(g')\, T(g)\mathbf{w}\rangle$$

$$= \frac{1}{|G|} \sum_{g' \text{ in } G} \langle T(g'g)\mathbf{v}, T(g'g)\mathbf{w}\rangle. \qquad (3.16)$$

But, for each fixed g in G, as g' ranges over the elements in G then so does $g'g$, and each element in G is included just once. Thus summing over g' in Eq. (3.16) is independent of g, and Eq. (3.16) can therefore also be written as

$$\langle T(g)\mathbf{v}, T(g)\mathbf{w}\rangle^\sim = \frac{1}{|G|} \sum_{g' \text{ in } G} \langle T(g')\mathbf{v}, T(g')\mathbf{w}\rangle$$

$$= \langle \mathbf{v}, \mathbf{w}\rangle^\sim. \qquad (3.17)$$

Thus $g \rightsquigarrow T(g)$ is unitary with respect to the scalar product defined in Eq. (3.15).

QED.

Corollary.

Any finite dimensional representation of a finite group is completely reducible into a direct sum of irreducible representations.

Note. Both the theorem and the corollary are also true for compact Lie groups.

3.4. IRREDUCIBLE REPRESENTATIONS

Given the theorems that show that, at least for certain types of group, certain classes of general representation are decomposable into a direct sum of irreducible representations, it is clear that a task of considerable importance is to find techniques for showing whether or not any specific representation is irreducible and to classify those that are. This is a vast subject in its own right and we can only begin to touch on it here.

First, we present a very well-known criterion for deciding if a representation is irreducible.

Theorem (Schur's Lemma).

If T_1 and T_2 are two irreducible representations of a group G, then every intertwining operator is either zero or an isomorphism of V_1 onto V_2 (in which case, the two representations are equivalent).

Proof.

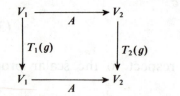

Let A be an intertwining operator between the two vector spaces V_1 and V_2 on which the two representations T_1 and T_2 are defined respectively. Thus

$$T_2(g)A = A\,T_1(g) \tag{4.1}$$

for all g in G.

(a) Let $\mathbf{v}$ belong to the kernel of A. Then Eq. (4.1) implies that, for all g,

$$A\,[T_1(g)\mathbf{v}] = T_2(g)\,[A\mathbf{v}] = 0$$

and so $T_1(g)$ maps Ker A into itself. But this is a subspace of V_1 and the representation T_1 is irreducible and hence has no proper G-invariant subspaces. Thus

$$\text{Ker } A = \{0\}, \text{ i.e., } A \text{ is one-to-one}$$

or

$$\text{Ker } A = V_1, \text{ i.e., } A \text{ is the trivial zero map.}$$

(b) Now, consider any element v_2 in the image of A, Im A. Thus there exists some v_1 in V_1 such that $v_2 = Av_1$. Then, Eq. (4.1) implies, for all g,

$$T_2(g)\, v_2 = T_2(g)\, Av_1 = A\,[T_1(g)\, v_1]$$

so that $T_2(g)v_2$ is also in the image of A. Therefore $T_2(g)$ maps Im A into itself. But T_2 is an irreducible representation and therefore V_2 has no proper G-invariant subspaces. Thus

$$\text{Im } A = \{0\}, \text{ i.e., } A \text{ is the zero map}$$

or

$$\text{Im } A = V_2, \text{ i.e., } A \text{ is surjective.}$$

Therefore, either A is the zero map or it is an isomorphism.

QED.

Corollary.

If T is an irreducible representation of a group G on a complex vector space V, then the only operator which commutes with all the operators $T(g)$ with g in G, is some complex multiple of the unit operator **1**.

Proof.

If, for all g in G, we have $AT(g) = T(g)A$ then, for every complex number μ,

$$(A - \mu\mathbf{1})\, T(g) = T(g)\, (A - \mu\mathbf{1}) \tag{4.2}$$

so that $A - \mu\mathbf{1}$ is an intertwining operator of the representation T with itself. But T is irreducible and hence, according to the theorem, either $A - \mu\mathbf{1}$ is invertible or else $A - \mu\mathbf{1} = 0$.

However, there always exists some complex number μ for which $A - \mu\mathbf{1}$ is *not* invertible. In the finite-dimensional case, this can be seen by mapping V isomorphically onto $\mathbb{C}^{\dim(V)}$ and then noting that the equation $\mathrm{Det}(A - \mu\mathbf{1}) = 0$ always has at least one solution μ [thinking of A now as a $\dim(V) \times \dim(V)$ matrix]; we shall prudently ignore the subtleties that arise when V has infinite dimension.

But then, if $A - \mu\mathbf{1}$ is not invertible for some complex number μ, the alternative must hold. Viz $A - \mu\mathbf{1} = 0$.

QED.

Note. This is one of the relatively rare situations when it really matters that V is a complex, not a real, vector space. In the real case, the equation $\mathrm{Det}(A - \mu\mathbf{1}) = 0$ may not have a solution (i.e., a real value for μ). This is for the same reason that the equation $x^2 + 1 = 0$ only has a solution if x is a *complex* number.

Now we come to an important collection of theorems concerning the unitary irreducible representations of a finite group. These culminate in the proof that every such representation is finite-dimensional and that, up to unitary equivalence, there are only a finite number of these representations. Many of these results have analogues in the case of compact Lie groups but they differ in significant details and the appropriate texts should be studied for further information.

Theorem.

If U is a unitary irreducible representation of a finite group G on a Hilbert space $\mathcal{H}$, then

(a) Dim $\mathcal{H} < \infty$,

(b) $\displaystyle\sum_{g \text{ in } G} \langle \mathbf{w}_1, U(g)\mathbf{v}_1 \rangle^* \langle \mathbf{w}_2, U(g)\mathbf{v}_2 \rangle = \frac{|G|}{\dim \mathcal{H}} \langle \mathbf{w}_1, \mathbf{w}_2 \rangle^* \langle \mathbf{v}_1, \mathbf{v}_2 \rangle$. (4.3)

Proof.

(a) The left hand side of Eq. (4.3) is a continuous and linear function of the vector $\mathbf{v}_2$ in $\mathcal{H}$ and hence equals $\langle \mathbf{k}(\mathbf{w}_1, \mathbf{w}_2, \mathbf{v}_1), \mathbf{v}_2 \rangle$ for some vector $\mathbf{k}$ in $\mathcal{H}$ that is itself a function of the vectors $\mathbf{w}_1$, $\mathbf{w}_2$ and $\mathbf{v}_1$.

[This uses the theorem that if $\mathbf{v} \rightsquigarrow F(\mathbf{v})$ is a continuous and linear complex-valued, function on the Hilbert space $\mathcal{H}$, then there exists some vector $\mathbf{k}$ such that $F(\mathbf{v}) = \langle \mathbf{k}, \mathbf{v} \rangle$ for all $\mathbf{v}$ in $\mathcal{H}$. The proof of this is relegated to one of the problems.]

The left hand side of Eq. (4.3) is also an antilinear function of $\mathbf{v}_1$ in $\mathcal{H}$ and hence (by the same argument) there exists some linear operator $A(\mathbf{w}_1, \mathbf{w}_2)$ such that $\mathbf{k}(\mathbf{w}_1, \mathbf{w}_2, \mathbf{v}_1) = A(\mathbf{w}_1, \mathbf{w}_2)\mathbf{v}_1$. Thus,

$$\sum_{g \text{ in } G} \langle \mathbf{w}_1, U(g)\mathbf{v}_1 \rangle^* \langle \mathbf{w}_2, U(g)\mathbf{v}_2 \rangle = \langle A(\mathbf{w}_1, \mathbf{w}_2)\mathbf{v}_1, \mathbf{v}_2 \rangle$$

and so, for all g' in G,

$$\langle A(\mathbf{w}_1, \mathbf{w}_2) U(g')\mathbf{v}_1, U(g')\mathbf{v}_2 \rangle$$

$$= \sum_{g \text{ in } G} \langle \mathbf{w}_1, U(gg')\mathbf{v}_1 \rangle^* \langle \mathbf{w}_2, U(gg')\mathbf{v}_2 \rangle .$$

But, for each fixed g', as g ranges over G so gg' also ranges over G and hence the sum is the same as

$$\sum_{g \text{ in } G} \langle \mathbf{w}_1, U(g)\mathbf{v}_1 \rangle^* \langle \mathbf{w}_2, U(g)\mathbf{v}_2 \rangle = \langle A(\mathbf{w}_1, \mathbf{w}_2)\mathbf{v}_1, \mathbf{v}_2 \rangle.$$

This is true for all $\mathbf{v}_1$, $\mathbf{v}_2$ in $\mathscr{H}$ and hence, for all g in G, we have $A(\mathbf{w}_1, \mathbf{w}_2) \, U(g) = U(g) \, A(\mathbf{w}_1, \mathbf{w}_2)$. But from the corollary to Schur's Lemma (with the proper proof for infinite-dimensional spaces), we then have $A(\mathbf{w}_1, \mathbf{w}_2) = \mu(\mathbf{w}_1, \mathbf{w}_2)\mathbf{1}$ for some complex number μ that is of course a function of the vectors $\mathbf{w}_1$ and $\mathbf{w}_2$. Thus,

$$\sum_{g \text{ in } G} \langle \mathbf{w}_1, U(g)\mathbf{v}_1 \rangle^* \langle \mathbf{w}_2, U(g)\mathbf{v}_2 \rangle = \mu(\mathbf{w}_1, \mathbf{w}_2)^* \langle \mathbf{v}_1, \mathbf{v}_2 \rangle. \quad (4.4)$$

We can repeat this line of argument and obtain the result that $\mu(\mathbf{w}_1, \mathbf{w}_2) = r\langle \mathbf{w}_1, \mathbf{w}_2 \rangle$ where r is a number which, by choosing $\mathbf{w}_1 = \mathbf{w}_2 = \mathbf{v}_1 = \mathbf{v}_2$, must be real. Thus we have shown that

$$\sum_{g \text{ in } G} \langle \mathbf{w}_1, U(g)\mathbf{v}_1 \rangle^* \langle \mathbf{w}_2, U(g)\mathbf{v}_2 \rangle = r\langle \mathbf{w}_1, \mathbf{w}_2 \rangle^* \langle \mathbf{v}_1, \mathbf{v}_2 \rangle \quad (4.5)$$

for some real number r.

(b) Now let $\{\mathbf{e}^1, \mathbf{e}^2, \dots, \mathbf{e}^N\}$ be an arbitrary finite orthonormal set of vectors in $\mathscr{H}$. Then Eq. (4.5) gives

$$\sum_{g \text{ in } G} |\langle \mathbf{e}^i, U(g)\mathbf{v} \rangle|^2 = r \| \mathbf{v} \|^2 \text{ for all } i = 1, \dots, N.$$

Thus,

$$\sum_{i=1}^{N} \sum_{g \text{ in } G} |\langle \mathbf{e}^i, U(g)\mathbf{v} \rangle|^2 = Nr \| \mathbf{v} \|^2. \quad (4.6)$$

But Bessel's inequality [Eq. (2.3.24)] shows that

$$\sum_{i=1}^{N} |\langle e^i, U(g)v\rangle|^2 \leq \| U(g)v \|^2 = \| v \|^2$$

and so,

$$\sum_{g \text{ in } G} \sum_{i=1}^{N} |\langle e^i, U(g)v\rangle|^2 \leq |G| \, \| v \|^2 . \qquad (4.7)$$

Since both sums in Eq. (4.6) are finite, we can interchange the order, which gives the left hand side of Eq. (4.7). Hence we conclude that

$$Nr \leq |G|. \qquad (4.8)$$

But $|G|$ is finite and $r > 0$, and so Eq. (4.8) implies that the largest possible value for N (which is dim $\mathscr{H}$) must be a finite number. Thus dim $\mathscr{H} < \infty$. Which proves the first part of the theorem.

(c) Now let $\{e^1, e^2, \ldots, e^n\}$ be an orthonormal *basis* for $\mathscr{H}$, where $n = \text{dim } \mathscr{H} < \infty$. Then the Bessel's inequality becomes an equality and so Eq. (4.8) gives $nr = |G|$. Hence

$$r = \frac{|G|}{\text{dim } \mathscr{H}}$$

which, when substituted into Eq. (4.5), gives the second assertion of the theorem in Eq. (4.3).

QED.

Theorem.

Let U_1 and U_2 be two inequivalent, unitary, irreducible representations of a finite group G defined on Hilbert spaces $\mathscr{H}_1$ and $\mathscr{H}_2$, respectively. Then, for all $\mathbf{v}_1$, $\mathbf{w}_1$ in $\mathscr{H}_1$ and $\mathbf{v}_2$, $\mathbf{w}_2$ in $\mathscr{H}_2$,

$$\sum_{g \text{ in } G} \langle \mathbf{w}_1, U_1(g)\mathbf{v}_1 \rangle^* \langle \mathbf{w}_2, U_2(g)\mathbf{v}_2 \rangle = 0 . \tag{4.9}$$

Proof.

Much as in the theorem above except that Schur's Lemma for inequivalent representations is used, rather than the corollary to the lemma for a single representation. (Exercise)

QED.

At this point, it is convenient to introduce the vector space $\mathscr{F} := \text{Map}(G, \mathbb{C})$ with the scalar product

$$\langle f_1, f_2 \rangle_{\mathscr{F}} := \frac{1}{|G|} \sum_{g \text{ in } G} f_1^*(g) f_2(g) . \tag{4.10}$$

(Exercise. Show that this does indeed define a scalar product on $\mathscr{F}$.) Then if we define the special class of functions

$$f_{\mathbf{w}\mathbf{v}}^{(\mu)}(g) := \langle \mathbf{w}, U_\mu(g)\mathbf{v} \rangle_\mu , \tag{4.11}$$

where μ denotes a particular irreducible representation U_μ of G defined on the Hilbert space $\mathscr{H}_\mu$ with scalar product $\langle \, , \, \rangle_\mu$, then Eqs. (4.3) and (4.9) can be written together in the succinct form

$$\langle f_{\mathbf{w}_1\mathbf{v}_1}^{(\mu)}, f_{\mathbf{w}_2\mathbf{v}_2}^{(\mu')} \rangle_{\mathscr{F}} = \frac{\delta^{\mu\mu'}}{\dim \mathscr{H}_\mu} \langle \mathbf{w}_1, \mathbf{w}_2 \rangle_\mu^* \langle \mathbf{v}_1, \mathbf{v}_2 \rangle_\mu . \tag{4.12}$$

Theorem.

The number of inequivalent irreducible representations of a finite group is finite. If the Hilbert spaces corresponding to these inequivalent representations are $\mathcal{H}_1, \mathcal{H}_2, \ldots, \mathcal{H}_N$ then

$$\sum_{\mu=1}^{N} [\dim \mathcal{H}_\mu]^2 \le |G|. \tag{4.13}$$

Proof.

In each Hilbert space $\mathcal{H}_\mu$, we can choose an orthonormal basis set $\{e^1, e^2, \ldots, e^n\}$ where $n = \dim \mathcal{H}_\mu$. Then Eq. (4.12) gives

$$\langle f^{(\mu)}_{e^i e^k}, f^{(\mu')}_{e^j e^l} \rangle_{\mathcal{F}} = \frac{\delta^{\mu\mu'}}{\dim \mathcal{H}_\mu} \delta_{ij} \delta_{kl}. \tag{4.14}$$

Equation (4.14) tells us that $\{ f^{(\mu)}_{e^i e^k}$, with $i, k = 1, \ldots, \dim \mathcal{H}_\mu\}$ is a set of $[\dim \mathcal{H}_\mu]^2$ orthonormal vectors in the complex vector space $\mathcal{F} = \mathrm{Map}(G, \mathbb{C})$. If we take any finite set of inequivalent, irreducible representations of G then each one contributes a set of $[\dim \mathcal{H}_\mu]^2$ vectors of this type and they are all orthogonal to each other. The total number of such vectors must, of course, be less that the dimension of $\mathrm{Map}(G, \mathbb{C})$. But $\dim [\mathrm{Map}(G, \mathbb{C})] = |G|$, which is finite since G is a finite group. We conclude that there can only be a finite number of inequivalent irreducible representations of G. And the total number of the vectors above being less than $|G|$ implies that Eq. (4.13) is satisfied.

QED.

Note. In the next section, we shall use a more refined analysis of the irreducible representations of a finite group to show that Eq. (4.13) is true with the inequality replaced with an equality. In this form, the result places very useful bounds on the number of in-equivalent irreducible representations of G and on their possible dimensions.

3.5. GROUP CHARACTERS

We will now introduce the concept of a 'character' of a group representation: an idea that is of considerable power in developing the general representation theory of a wide class of groups. First a preliminary definition.

Definition.

The *trace* of a linear operator A on a finite-dimensional Hilbert space $\mathcal{H}$ is defined as

$$\text{Tr } A := \sum_{i=1}^{n} \langle \mathbf{e}^i, A\mathbf{e}^i \rangle, \tag{5.1}$$

where $\{\mathbf{e}^1, \mathbf{e}^2, \ldots, \mathbf{e}^n\}$ is an orthonormal basis for $\mathcal{H}$ with $n = \dim \mathcal{H}$.

Note. (1) For this definition to make sense, it is necessary that the sum in Eq. (5.1) be independent of the particular basis that is chosen. To see that this is indeed the case let $\{\mathbf{e}^{1'}, \mathbf{e}^{2'}, \ldots, \mathbf{e}^{n'}\}$ be any other orthonormal basis set. Then, using Eq. (2.3.27), we may write $\text{Tr } A$ as

$$\text{Tr } A = \sum_{i=1}^{n} \sum_{j=1}^{n} \langle \mathbf{e}^i, \mathbf{e}^{j'} \rangle \langle \mathbf{e}^{j'}, A\mathbf{e}^i \rangle$$

$$= \sum_{i=1}^{n} \sum_{j=1}^{n} \langle \mathbf{e}^i, \mathbf{e}^{j'} \rangle \langle A^\dagger \mathbf{e}^{j'}, \mathbf{e}^i \rangle$$

$$= \sum_{j=1} \sum_{i=1} \langle A^\dagger e^{j'}, e^i \rangle \langle e^i, e^{j'} \rangle$$

$$= \sum_{j=1} \langle A^\dagger e^{j'}, e^{j'} \rangle \quad \text{[on using Eq. (2.3.27) again]}$$

$$= \sum_{j=1} \langle e^{j'}, A e^{j'} \rangle$$

which proves the basis independence of the definition.

(2) On the Hilbert space $\mathcal{H} = \mathbb{C}^n$, Tr A is just the usual matrix trace

$$\text{Tr } A = \sum_{i=1}^{n} A_{ii}.$$

(3) For any pair of operators A and B, we have (Exercise)

$$\text{Tr } (AB) = \text{Tr } (BA)$$

$$\text{Tr } (aA + bB) = a\text{Tr } A + b \text{ Tr } B,$$

(5.2)

where a and b are any pair of complex numbers.

Definition.

Let U be a unitary representation of a group G on a finite-dimension Hilbert space $\mathcal{H}$. Then the *character* of U is the map $\chi : G \to \mathbb{C}$ defined by:

$$\chi(g) := \text{Tr } U(g).$$

(5.3)

Note. (1) The character of a representation is invariant under similarity transformations since

$$\text{Tr}\,[AU(g)\,A^{-1}] = \text{Tr}\,U(g) \qquad (5.4)$$

by virtue of Eq. (5.2). Thus equivalent representations have the same character.

(2) For any g_0 in G, we have

$$\chi(g_0 g g_0^{-1}) = \text{Tr}\,U(g_0 g g_0^{-1}) = \text{Tr}\,[U(g_0)\,U(g)\,U(g_0)^{-1}]$$

$$= \text{Tr}\,U(g)$$

and so the character is constant on the conjugacy classes of G which were defined in Sec. 1.4 as the orbits of the left action of G on itself given by Eq. (1.4.8).

(3) At the identity e in G, we have

$$\chi(e) = \text{Tr}\,\mathbf{1} = \dim \mathcal{H}. \qquad (5.5)$$

Theorem.

Using the notation of the previous section, the characters χ_μ of the set of inequivalent irreducible representations $U_1, U_2, \ldots, U_N$ defined on Hilbert spaces $\mathcal{H}_1, \mathcal{H}_2, \ldots, \mathcal{H}_N$, are orthonormal functions in the vector space $\mathcal{F} = \text{Map}(G, \mathbb{C})$ equipped with the scalar product in Eq. (4.10), i.e.,

$$\langle \chi_\mu, \chi_{\mu'} \rangle_\mathcal{F} = \delta_{\mu\mu'} . \qquad (5.6)$$

Proof.

$$\chi_\mu(g) = \sum_{i=1}^{\dim \mathcal{H}_\mu} \langle e^i, U_\mu(g) e^i \rangle = \sum_{i=1}^{\dim \mathcal{H}_\mu} f_{e^i e^i}^{(\mu)}(g) \quad [\text{cf. Eq. (4.11)}].$$

The theorem then follows by putting $i = k$ and $j = l$ in Eq. (4.14) and summing over i and k from $1, \ldots, \dim \mathcal{H}_\mu$.

QED.

It follows from this result that inequivalent irreducible representations have different characters, and so we arrive at the fundamental conclusion:

> The equivalence class of an irreducible representation of a finite group G is uniquely specified by its character.

Characters can also be used to study the properties of reducible representations of a group that can be decomposed into a direct sum of irreducible representations in the way discussed in Sec. 3.3. In particular, the character of a direct sum $U_1 \oplus U_2$ of representations U_1 and U_2 is related to the characters of its summands by

$$\chi_{U_1 \oplus U_2} = \chi_{U_1} + \chi_{U_2}. \qquad (5.7)$$

Suppose then that U is any finite-dimensional representation of a finite group G. We know from the theorem in Sec. 3.3 that every such representation is completely reducible and hence it can be decomposed into a direct sum of irreducible representations in the form

$$U = n_1 U_1 \oplus n_2 U_2 \oplus \ldots \oplus n_N U_N, \qquad (5.8)$$

where $U_1, U_2, \ldots, U_N$ are representatives of the finite number N of inequivalent irreducible representations of the finite group G and $n_i U_{n_i}$ means that we have n_i copies of U_{n_i} in the direct sum decomposition of U. It follows from Eq. (5.7) that the character χ_U can be written uniquely as

$$\chi_U(g) = \sum_{\mu=1}^{N} n_\mu \chi_\mu(g). \qquad (5.9)$$

It then follows from Eq. (5.6) that the number of times, n_μ, that a particular representation U_μ occurs in the direct sum decomposition of U can be computed directly in terms of the characters of U and U_μ as

$$n_\mu = \langle \chi_\mu, \chi_U \rangle = \frac{1}{|G|} \sum_{g \text{ in } G} \chi_\mu^*(g) \chi_U(g). \qquad (5.10)$$

Now we can prove the promised refinement of Eq. (4.13).

Theorem.

The dimensions of the inequivalent irreducible representations of the finite group G satisfy the equation

$$\sum_{\mu=1}^{N} [\dim \mathscr{H}_\mu]^2 = |G|. \qquad (5.11)$$

Proof.

In proving this theorem we introduce a concept that is of considerable interest in its own right. This is the *regular representation* of the group G which is defined on the vector space $\mathscr{F} = \text{Map}(G, \mathbb{C})$ by

$$[R(g)f](g') := f(g^{-1}g'). \qquad (5.12)$$

Note that this representation $g \rightsquigarrow R(g)$ is unitary with respect to the inner product on $\text{Map}(G, \mathbb{C})$ defined in Eq. (4.10).

A useful orthonormal basis for the $|G|$-dimensional vector space $\text{Map}(G, \mathbb{C})$ is the set of functions $\{f^{g'}, \text{ for } g' \text{ in } G\}$ where

$$f^{g'}(g) := |G|^{\frac{1}{2}} \quad \text{if } g = g',$$
$$:= 0 \qquad \text{otherwise.} \qquad (5.13)$$

In terms of this basis set, the character of the regular representation is

$$\chi_R(g) = \sum_{g'} \langle f^{g'}, R(g)f^{g'} \rangle = \sum_{g'} \sum_{g''} f^{g'}(g'') f^{g'}(g^{-1}g'') \frac{1}{|G|}$$

$$= \sum_{g'} f^{g'}(g^{-1}g') \frac{1}{|G|^{\frac{1}{2}}},$$

where the sums are over the set of group elements in G. Thus

$$\chi_R(g) = |G| \quad \text{if } g = e, \qquad (5.14)$$
$$= 0 \quad \text{otherwise.}$$

Let m_μ be the number of times that the irreducible representation U_μ occurs in the direct sum decomposition of the regular representation. By Eq. (5.10), we have

$$m_\mu = \langle \chi_\mu, \chi_R \rangle = \frac{1}{|G|} \sum_g \chi_\mu^*(g) \chi_R(g)$$

$$= \chi_\mu^*(e) = \dim \mathscr{H}_\mu \text{ by Eq. (5.5).} \qquad (5.15)$$

Hence, the important result:

> The number of times that a particular representation U_μ occurs in the direct sum decomposition of the regular representation is equal to the dimension of the representation.

In general, the dimension of a direct sum representation is

$$\dim [n_1 U_1 \oplus n_2 U_2 \oplus \ldots \oplus n_N U_N] = \sum_{\mu=1}^{N} n_\mu [\dim \mathscr{H}_\mu] \quad (5.16)$$

so, for the regular representation,

$$\dim(R) = \sum_{\mu=1}^{N} [\dim \mathscr{H}_\mu]^2 . \quad (5.17)$$

But, $|G| = \dim [\text{Map}(G, \mathbb{C})]$, which proves the result.

QED.

3.6. THE GROUP ALGEBRA

The final concept to be introduced in this lecture course (and another very powerful one it is too) is that of the 'group algebra' of a finite group G. To motivate the precise definition, let us first consider an arbitrary representation T of a finite group on a vector space V.

If f belongs to $\text{Map}(G, \mathbb{C})$ define the linear operator $T(f)$ on V by

$$T(f) := \sum_{g} f(g) \, T(g) . \quad (6.1)$$

Then,

$$T(f_1) \, T(f_2) = \sum_{g} \sum_{g'} f_1(g) f_2(g') \, T(gg')$$

$$= \sum_{g} \sum_{g^{-1}g''} f_1(g) f_2(g^{-1}g'') \, T(g'') .$$

But summing over $g^{-1}g''$ is equivalent to summing over g'' (for each fixed g). Therefore,

$$T(f_1)\,T(f_2) = \sum_{g'}\left[\sum_{g} f_1(g)\,f_2(g^{-1}g')\right]T(g') \qquad (6.2)$$

which motivates the following definition.

Definition.

The *group algebra* $K(G)$ associated with a finite group G is the vector space Map$(G, \mathbb{C})$ equipped with the additional 'product law':

$$[f_1*f_2](g') := \sum_{g} f_1(g)\,f_2(g^{-1}g') . \qquad (6.3)$$

The 'product' f_1*f_2 of two elements f_1, f_2 in Map$(G, \mathbb{C})$ is known as the *convolution* of f_1 and f_2.

Note. (1) The convolution product is associative since

$$f_1*(f_2*f_3) = (f_1*f_2)*f_3 \qquad (6.4)$$

for all f_1, f_2 and f_3 in Map$(G, \mathbb{C})$. (Exercise)
(2) The basis vectors f^g with g in G, defined in Eq. (5.13), have the property that

$$[f^g*f^{g'}](g_0) = \sum_{g''} f^g(g'')\,f^{g'}(g''^{-1}g_0) = f^{g'}(g^{-1}g_0)$$

$$= 1 \quad \text{if } g^{-1}g_0 = g'$$

$$= 0 \quad \text{otherwise.}$$

Therefore,

$$f^g * f^{g'} = f^{gg'} \quad \text{for all } g \text{ and } g' \text{ in } G. \qquad (6.5)$$

(3) Equation (6.2) shows that every representation of G gives rise to a representation of the algebra $K(G)$, i.e.,

$$T(f_1)\, T(f_2) = T(f_1 * f_2). \qquad (6.6)$$

Conversely, let $f \rightsquigarrow T(f)$ be any representation of the algebra $K(G)$, i.e., a set of operators on some vector space V satisfying Eq. (6.6) and being linear in f. Define

$$T(g) := T(f^g), \qquad (6.7)$$

where, as before, the basis vector f^g in Map$(G, \mathbb{C})$ is defined as in Eq. (5.13). Then, for all g_1 and g_2 in G,

$$T(g_1)\, T(g_2) = T(f^{g_1})\, T(f^{g_2}) = T(f^{g_1} * f^{g_2})$$

$$= T(f^{g_1 g_2}) \quad \text{by Eq. (6.5)}$$

$$= T(g_1 g_2)$$

which is therefore a representation of G.
We conclude therefore that:

> The representations of the group algebra $K(G)$ and the group G are in one-to-one correspondence.

The deep significance of this result is that algebras like $K(G)$ have a well-developed representation theory of their own and this may therefore be used to study the representations of the group. Our use of $K(G)$ will be relatively mild but it will enable us to augment the information contained in Eq. (5.11) with a way of computing the value of N, i.e., the number of inequivalent unitary irreducible representations of the finite group G. But first, some rather technical looking results.

Lemma.

Let U be an irreducible representation of G on a finite-dimensional Hilbert space $\mathscr{H}$. Then any linear operator A on $\mathscr{H}$ can be written as $A = U(f_A)$ for some f_A in $\mathrm{Map}(G, \mathbb{C})$.

Proof.

Given A, we define the function

$$f_A(g) := \frac{(\dim \mathscr{H})}{|G|} \, \mathrm{Tr}\,[U^\dagger(g)\,A]. \qquad (6.8)$$

Then, for any pair of vectors $\mathbf{v}$, $\mathbf{w}$ in $\mathscr{H}$,

$$\langle \mathbf{v}, U(f_A)\mathbf{w}\rangle = \sum_g f_A(g)\,\langle \mathbf{v}, U(g)\mathbf{w}\rangle .$$

Therefore,

$$\langle \mathbf{v}, U(f_A)\mathbf{w}\rangle = \sum_g \sum_{i=1}^{\dim \mathscr{H}} \langle \mathbf{e}^i, U^\dagger(g)\,A\mathbf{e}^i\rangle\langle \mathbf{w}, U^\dagger(g)\mathbf{v}\rangle^* \frac{\dim \mathscr{H}}{|G|}$$

$$= \sum_{i=1}^{\dim \mathscr{H}} \sum_g \langle A\mathbf{e}^i, U(g)\,\mathbf{e}^i\rangle^*\langle \mathbf{v}, U(g)\mathbf{w}\rangle \frac{\dim \mathscr{H}}{|G|}$$

$$= \sum_{i=1}^{\dim \mathcal{H}} \langle \mathbf{v}, A\mathbf{e}^i \rangle \langle \mathbf{e}^i, \mathbf{w} \rangle \text{ [from Eq. (4.3)]}$$

$$= \sum_{i=1}^{\dim \mathcal{H}} \langle A^\dagger \mathbf{v}, \mathbf{e}^i \rangle \langle \mathbf{e}^i, \mathbf{w} \rangle = \langle A^\dagger \mathbf{v}, \mathbf{w} \rangle = \langle \mathbf{v}, A\mathbf{w} \rangle$$

which, since it is true for all $\mathbf{v}$ and $\mathbf{w}$ in $\mathcal{H}$, implies that $U(f_A) = A$.

QED.

Theorem.

    Let $U_1, U_2, \ldots, U_N$ be representatives of the set of all unitary inequivalent, irreducible representations of the finite group G and let $\mathcal{H}_1, \mathcal{H}_2, \ldots, \mathcal{H}_N$ be the Hilbert spaces on which they are defined. Then the map

$$j : K(G) \rightarrow B(\mathcal{H}_1) \oplus B(\mathcal{H}_2) \oplus \ldots \oplus B(\mathcal{H}_N)$$

defined by

$$j(f) := U_1(f) \oplus U_2(f) \oplus \ldots \oplus U_N(f) \tag{6.9}$$

is an isomorphism of both the vector space structure and the multiplicative monoid structure on the two spaces.

[$B(\mathcal{H})$ denotes the set of all operators on the finite-dimensional Hilbert space $\mathcal{H}$. The 'multiplicative monoid' structure refers to operator product defined in Eq. (2.4.3).]

Proof.

(a) The definition of $T(f)$ in Eq. (6.1) shows that the map j is linear and Eq. (6.6) shows that it preserves the product structures on the two spaces; i.e., it is a morphism.

(b) The Lemma above shows that the map $j_\mu : K(G) \rightarrow B(\mathcal{H}_\mu)$

defined by $j_\mu(f) := U_\mu(f)$ is surjective. Hence j is also surjective.

(c) It remains to show that j is injective. So, let f belong to Ker j. This implies that $\sum_g f(g) U_\mu(g)$ must vanish for each of the representations U_μ of G. But every finite-dimensional representation U of G decomposes into a direct sum of finite multiples of this basic set $U_1, U_2, \ldots, U_N$. Hence, for every such representation U,

$$\sum_g f(g) U(g) = 0. \qquad (6.10)$$

But, there is a one-to-one correspondence between the representations of G and those of $K(G)$. Hence Eq. (6.10) implies that we must have $U(f) = 0$ for every representation of $K(G)$. In particular, this must be true for the representation of $K(G)$ on itself defined by

$$[R(f)](f') := f * f'. \qquad (6.11)$$

But $R(f)f' = f * f' = 0$ for all f' in $K(G)$ implies in particular that this must be true for the basis functions $f^{g'}$ defined in Eq. (5.13). However, $f * f^{g'}(g) = f(gg'^{-1})$ and so $f * f^{g'}(e) = f(g'^{-1})$, and f in Ker j implies that f vanishes on every g'^{-1} for g' in G. But this set is all of G, and hence f vanishes.

Hence, Ker j is trivial and so j is an injection. Which completes the proof of the theorem.

QED.

Now we can derive the promised result about the number of inequivalent irreducible representations of G.

Theorem.

The number N of equivalence classes of inequivalent, unitary, irreducible representations of G is equal to the number of conjugacy classes of G.

Proof.

(a) Let f_0 belongs to the *centre* of $K(G)$, i.e., it commutes with all members of $K(G)$

$$f_0*f = f*f_0 \text{ for all } f \text{ in } K(G) . \qquad (6.12)$$

From the theorem, we have that $K(G)$ is isomorphic to the direct sum $B(\mathcal{H}_1) \oplus B(\mathcal{H}_2) \oplus \ldots \oplus B(\mathcal{H}_N)$ and hence, in particular, both algebras must have the same centre. But for an operator to belong to the centre of $B(\mathcal{H})$ means that it commutes with every operator on $\mathcal{H}$ and, using the Corollary to Schur's Lemma applied to the group $B(\mathcal{H})$ represented on $\mathcal{H}$ with itself, this implies that the centre of $B(\mathcal{H})$ is a multiple of the unit operator for any finite-dimensional Hilbert space $\mathcal{H}$.

In particular, this applies to each $B(\mathcal{H}_\mu)$ and hence the centre of the direct sum is all operators of the form $(a_1\mathbf{1}_1, a_2\mathbf{1}_2, \ldots, a_N\mathbf{1}_N)$, where $a_1, a_2, \ldots, a_N$ are arbitrary complex numbers and $\mathbf{1}_\mu$ refers to the unit operator on $\mathcal{H}_\mu$. The set of all operators of this form is clearly a subspace of the set of all operators in the direct sum and has dimension N. Thus

$$\dim [\text{centre of } K(G)] = N. \qquad (6.13)$$

(b) On the other hand, we can also study the centre of $K(G)$ directly. Indeed, suppose that f_0 belongs to this centre so that Eq. (6.12) is satisfied for all f in $K(G)$. Then, since every f in $K(G)$ can be written identically as

$$f = \sum_g f(g) f^g,$$

(6.14)

where $\{f^g, g \text{ in } G\}$ is the basis set defined in Eq. (5.13), it follows that f_0 belongs to the centre of $K(G)$ if, and only if,

$$f^g * f_0 = f_0 * f^g$$

(6.15)

for all g in G.
But,

$$f^g * f_0(g') = \sum_{g''} f^g(g'') f_0(g''^{-1} g') = f_0(g^{-1} g') \quad (6.16)$$

and,

$$f_0 * f^g(g') = \sum_{g''} f_0(g'') f^g(g''^{-1} g') = f_0(g' g^{-1}). \quad (6.17)$$

This means that f_0 is in the centre of $K(G)$ if and only if, for each g,

$$f_0(g^{-1} g') = f_0(g' g^{-1}) \text{ for all } g' \text{ in } G$$

which is equivalent to

$$f_0(gg' g^{-1}) = f_0(g') \text{ for all } g \text{ and } g' \text{ in } G$$

(6.18)

and this is true if and only if f_0 is constant on the conjugacy classes of G. Thus this is a necessary and sufficient condition for f_0 to lie in the centre of $K(G)$. The dimension of $K(G)$ is thus equal to the number of conjugacy classes and this, together with Eq. (6.13), proves the theorem.

QED.

Note. This theorem and the theorem involved in the proof of Eq. (5.11) together make up what is known as *Burnside's theorem*:

(1) The number of equivalence classes of unitary, irreducible representations of a finite group G is equal to the number of conjugacy classes of G.
(2) The dimensions of these representations satisfy the relation

$$\sum_{\mu=1}^{N} [\dim \mathscr{H}_\mu]^2 = |G|.\tag{6.19}$$

This is a powerful result from which much useful information can be extracted.

Examples.

(1) Let G be a finite abelian group. Then each conjugacy class consists of a single group element and hence the number of conjugacy classes is equal to $|G|$. But each irreducible representation satisfies the inequality $\dim \mathscr{H}_\mu \geq 1$ and then Eq. (6.19) implies that $\dim \mathscr{H}_\mu = 1$. So we have derived the important result:

> Every irreducible representation of a finite abelian group is one-dimensional.

(2) The symmetric group S_3 has 6 elements in 3 conjugacy classes:

$$\{e\}, \{231, 312\} \text{ and } \{132, 213, 321\},$$

where the (rather non-standard) notation abc means the permutation map

$$\begin{bmatrix} 1 \rightarrow a \\ 2 \rightarrow b \\ 3 \rightarrow c \end{bmatrix}$$

Thus there are three inequivalent classes of irreducible representation, and the dimensions of the associated Hilbert spaces satisfy the relation (6.19)

$$[\dim \mathcal{H}_1]^2 + [\dim \mathcal{H}_2]^2 + [\dim \mathcal{H}_3]^2 = 6 . \qquad (6.20)$$

The only possible choices for the dimensions which satisfy this equation are the three integers 1, 1, 2 (i.e., $1^2 + 1^2 + 2^2 = 6$) and so we can conclude that the group S_3 has two inequivalent one-dimensional irreducible representations and a single two-dimensional irreducible representation.

PROBLEMS

1. Show that the order of the symmetric group S_N is $|S_N| = N!$

2. The Cartesian product $G_1 \times G_2$ of a pair of groups G_1 and G_2 is defined to be the set of all pairs (g_1, g_2) with $g_1 \in G_1$ and $g_2 \in G_2$. Show that it can be given a group structure in which the law of combination is

$$(g_1, g_2)(g_1', g_2') := (g_1 g_1', g_2 g_2').$$

Prove that $|G_1 \times G_2| = |G_1| \times |G_2|$.

3. What are the proper subgroups of $\mathbb{Z}_4 = \{e, a, a^2, a^3\}$ other than $\mathbb{Z}_2 := \{e, a^2\}$?

4. If $Y \subset X$ define $j : \mathrm{Perm}(Y) \to \mathrm{Perm}(X)$ by

$$(j(f))(x) := f(x) \text{ if } x \in Y \subset X$$
$$:= x \quad \text{otherwise.}$$

Show that j is an isomorphism of $\mathrm{Perm}(Y)$ onto a subgroup of $\mathrm{Perm}(X)$.

5. Show that the real dimension of the Lie group $U(n)$ is n^2 and that $SU(n)$ has dimension $n^2 - 1$.

6. Let $\mu : G_1 \to G_2$ be a homomorphism so that $\mu(g_1 g_2) = \mu(g_1) \mu(g_2)$ for all $g_1, g_2 \in G_1$. Show that $\mu(e_1) = e_2$ and that $\mu(g^{-1}) = (\mu(g))^{-1}$ for all $g \in G_1$.

7. Show that if G acts on the left of a set X, then $G_{x_0} = \{g \in G \mid gx_0 = x_0\}$ is a subgroup of G for any $x_0 \in X$.

8. If $\mu : G_1 \to G_2$ is a homomorphism, show that Ker μ and Im μ are subgroups of G_1 and G_2 respectively.

9. Let μ be the map from the Lie group SL(2, $\mathbb{C}$) to the group of Möbius transformations of the complex plane which associates with $\begin{pmatrix} a & b \\ c & d \end{pmatrix} \in$ SL(2, $\mathbb{C}$) the transformation $z \to \dfrac{az+b}{cz+d}$. Show that μ is a homomorphism.

10. (a) Show that a group G satisfies the 'cancellation laws':
    (i) If a, b, c, d in G are such that $ab = cb$, then $a = c$.
    (ii) If a, b, c, d in G are such that $ba = bc$, then $a = c$.

    (b) Show that in a group G,

    $$(ab)^{-1} = b^{-1}a^{-1}$$

    for all a and b in G.

11. If G is any group and X is any set, show that the set Map(X, G) of all maps of X into G has a group structure in which

    $$(f_1 f_2)(x) := f_1(x) f_2(x) \quad \text{for all } x \text{ in } X,$$

    is the combination law for the maps f_1 and f_2.

12. Show that the set of all Möbius transformations [Eq. (1.3.1)] of the complex plane forms a subgroup of the set of all transformations.

13. Construct explicitly the group table for all possible groups of order 4 and hence show that the only possibilities are $\mathbb{Z}_4$ and $\mathbb{Z}_2 \times \mathbb{Z}_2$.

14. Construct the group table for the group S_3 of order 6 and show that this group can be represented by the set $\{e, x, y, y^2, xy, xy^2\}$ of six elements subject to the relations

$$x^2 = y^3 = e, \ yx = xy^2, \ y^2x = xy.$$

Show that this group is non-abelian by finding a pair of elements a and b such that $ab \neq ba$.

15. Show that the set of all 'triangular' matrices $\begin{pmatrix} a & c \\ 0 & b \end{pmatrix}$ with a, b, c arbitrary complex numbers, subject to the constraint $ab \neq 0$, is a subgroup of the Lie group $GL(2, \mathbb{C})$.

16. Show that the set of all automorphisms of a group G is itself a group.

    Calculate the complete set of automorphisms for the cases $G = \mathbb{Z}_2$ and $G = \mathbb{Z}_4$ and show by direct calculation in both cases that the automorphisms do indeed form a group. Identify the group in both cases.

17. (a) If H and K are subgroups of a group G show that $H \cap K$ is also a subgroup of G.

    (b) Find an explicit example in the group S_3 (Q.14) of a pair of subgroups H and K for which the set $H \cup K$ is not a subgroup.

    (c) If H and K are subgroups of a group G define HK as

    $$HK := \{hk \text{ in } G \text{ with } h \text{ in } H \text{ and } k \text{ in } K\}.$$

    Show that HK is a subgroup of G if and only if $HK = KH$.

(d) Prove that a non-empty set $H \subset G$ is a subgroup of G if and only if ab^{-1} belongs to H whenever a and b belong to H.

18. (a) Two subgroups H_1 and H_2 of a group G are said to be *conjugate* if there exists some $g \in G$ such that

$$H_2 = gH_1g^{-1} := \{ghg^{-1} \text{ with } h \in H\}.$$

Show that in such a pair of subgroups H_1 is isomorphic to H_2.

(b) Find a pair of $\mathbb{Z}_2$ subgroups of the symmetric group S_3 that are conjugate.

19. If $l_g : X \to X$ is a left action of some group on a set X, show that a right action of G on X is given by $r_g(x) := l_g^{-1}(x)$.

20. (a) If a group acts on a set X, show that two orbits O_{x_1} and O_{x_2} (where x_1 and x_2 are points in X) either coincide or are disjoint.

(b) Show that the stability/isotropy groups G_{x_1} and G_{x_2} at any two points lying in the same orbit are conjugate.

21. The *centre* of a group G is defined to be the set of elements of G which commute with all elements of G, i.e.,

$$C(G) := \{g \in G \text{ such that } gg' = g'g \text{ for all } g' \in G\}.$$

Show that the centre $C(G)$ is a normal subgroup of G.

22. Find the centre of the Lie group SU2.

23. With reference to Q.15, show that the set of all triangular matrices of the form $\begin{pmatrix} 1 & d \\ 0 & 1 \end{pmatrix}$ with d an arbitrary complex number is a normal subgroup of the group considered in that question.

24. Show that the map $\mu : \mathbb{R} \to \mathbb{R}_+$ defined by $\mu(x) := e^x$ is a homomorphism of the additive group of the real line onto the multiplicative group of positive real numbers. What is its kernel?

25. Construct a homomorphism from the additive group $\mathbb{R}$ of the real numbers onto the group $U(1)$. Hence show that $U(1) \cong \mathbb{R}/\mathbb{Z}$.

26. Let X be a set with $N < \infty$ elements.

    (a) Construct an isomorphism between the complex vector spaces $\mathbb{C}^N$ and Map $(X, \mathbb{C})$.

    (b) A basis set for $\mathbb{C}^N$ is $\{(1, 0, \ldots, 0), (0, 1, \ldots, 0), \ldots, (0, 0, \ldots, 1)\}$. What is the corresponding basis induced in Map $(X, \mathbb{C})$ by this isomorphism?

    (c) Show that Map $(X, \mathbb{C})$ can be given a scalar product by defining:

    $$\langle f_1, f_2 \rangle := \sum_{x \in X} f_1^*(x) f_2(x).$$

    (d) How is this scalar product related to the usual one on $\mathbb{C}^N$ via the isomorphism in part (a)?

27. Construct an explicit isomorphism between the vector spaces $\mathbb{C}^{n^2}$ and $M(n, \mathbb{C})$.
    What is the scalar product on $M(n, \mathbb{C})$ induced from the usual one on $\mathbb{C}^{n^2}$ by this isomorphism? (It is sufficient, and probably more instructive, to do just the case $n=2$ in detail.)

28. Any complex vector space V can be regarded as a real vector space $V_{\mathbb{R}}$ by defining this to be the same set as V and with the addition of a pair of vectors being the same as for V but with scalar multiplication restricted to the real numbers.

(a) If $\dim(V) = n < \infty$ show that the real dimension of $V_{\mathbb{R}}$ is $2n$.

(b) Construct an explicit isomorphism between the real vector space $\mathbb{C}_{\mathbb{R}}^n$ and $\mathbb{R}^{2n}$.

(c) Show that the set of hermitian $n \times n$ matrices (i.e., $A^\dagger = A$, or equivalently, $A_{ij}^* = A_{ji}$ for all $i, j = 1, \ldots, n$) is a linear subspace of the real vector space $M(n, \mathbb{C})_{\mathbb{R}}$ but not of the complex vector space $M(n, \mathbb{C})$.

29. If V is a complex vector space, a function $F: V \to \mathbb{C}$ is said to be *linear* if, for all $\mu_1, \mu_2 \in \mathbb{C}$ and $\mathbf{v}_1, \mathbf{v}_2 \in V$, we have

$$F(\mu_1 \mathbf{v}_1 + \mu_1 \mathbf{v}_2) = \mu_1 F(\mathbf{v}_1) + \mu_2 F(\mathbf{v}_2),$$

and the set of all such linear functions is called the (algebraic) *dual* of V and is denoted V^*.

(a) Show that V^* may be given the structure of a complex vector space.

(b) If $\mathscr{H}$ is a finite-dimensional Hilbert space, use an orthonormal basis set $\{\mathbf{e}^1, \mathbf{e}^2, \ldots, \mathbf{e}^n\}$, where $n = \dim \mathscr{H}$, to show that every $F \in \mathscr{H}^*$ can be written in the form

$$F(\mathbf{v}) = \langle \mathbf{w}_F, \mathbf{v} \rangle \qquad \text{for all } \mathbf{v} \in \mathscr{H},$$

for some unique vector $\mathbf{w}_F$ in $\mathscr{H}$.
Hence, deduce that $\mathscr{H}^*$ and $\mathscr{H}$ are isomorphic.

30. In writing $\mathbf{v}^N \to \mathbf{v}$ for the strong convergence of a sequence $\mathbf{v}^N$ of vectors in a normed vector space V, it is implicitly presumed that there is only *one* limit vector, namely $\mathbf{v}$.
Show that this is indeed true, viz. if $\mathbf{v}^N \to \mathbf{v}$ and $\mathbf{v}^N \to \mathbf{v}'$ simultaneously then $\mathbf{v} = \mathbf{v}'$. (Hint: Study $\|\mathbf{v} - \mathbf{v}'\|$.)

31. Let $\mathcal{H}$ be an infinite-dimensional, separable Hilbert space with an orthonormal basis $\{e^1, e^2, \ldots\}$

   (a) Show that the infinite sum $\displaystyle\sum_{i=1}^{\infty} \mu_i e^i$ exists if and only if the sequence of complex numbers $\mu_1, \mu_2, \ldots$ satisfies

   $$\sum_{i=1}^{\infty} |\mu_i|^2 < \infty.$$

   (b) Hence construct an isomorphism between the Hilbert space ℓ_2 and any other infinite-dimensional, separable Hilbert space $\mathcal{H}$. (**Note.** This implies in particular that *any* two such Hilbert spaces $\mathcal{H}_1$ and $\mathcal{H}_2$ are isomorphic.)

32. A linear function $F : \mathcal{H} \to \mathbb{C}$, where $\mathcal{H}$ is an infinite-dimensional, separable Hilbert space, is said to be *continuous* if

   $$\lim_{N \to \infty} F(v^N) = F(v)$$

   for any strongly convergent sequence of vectors $v^N \to v$.

   Complete the work began in Q.29 by showing that any continuous linear function $F : \mathcal{H} \to \mathbb{C}$ can be written in the form

   $$F(v) = \langle w_F, v \rangle \qquad \text{for all } v \in \mathcal{H}$$

   for some unique vector $w_F \in \mathcal{H}$.

33. (a) Prove the adjoint $A^\dagger$ of a bounded operator A on a Hilbert space $\mathcal{H}$ is also bounded and that $\|A^\dagger\| = \|A\|$.

   (b) If A and B are bounded operators on $\mathcal{H}$ show that the product AB is also bounded and that $\|AB\| \le \|A\| \|B\|$.

34. A bounded operator on a Hilbert space $\mathscr{H}$ is a *projection operator* if

$$P^\dagger = P \quad \text{and} \quad P = P^2.$$

(a) Show that $(\mathbb{1}-P)$ is also a projection operator.

(b) Show that $\|P\| \le 1$.

(c) Find the eigenvalues of P and hence show that $\|P\| = 1$.
(Hint: Consider the implications of the equation $P^2 = P$.)

(d) In the direct sum decomposition of $\mathscr{H}$ with respect to a closed subspace W, $\mathscr{H} \cong W \oplus W_\perp$, a vector $\mathbf{v} \in \mathscr{H}$ can be decomposed uniquely in the form $\mathbf{v} = \mathbf{v}_W + \mathbf{v}_{W_\perp}$ with $\mathbf{v}_W \in W$ and $\mathbf{v}_{W_\perp} \in W_\perp$. Show that the linear operator defined on $\mathscr{H}$ by $P(\mathbf{v}) := \mathbf{v}_W$ is a projection operator.

35. Let $\{\mathbf{e}^1, \mathbf{e}^2, \ldots, \mathbf{e}^n\}$ and $\{\mathbf{f}^1, \mathbf{f}^2, \ldots, \mathbf{f}^n\}$ be two sets of basis vectors for a finite-dimensional complex vector space V. Let A be a linear operator on V and let M and N denote the matrix representatives of A with respect to the two bases.

Show that M and N are related by a similar transformation. That is, there is some $n \times n$ invertible matrix B such that

$$M = BNB^{-1}.$$

36. With reference to Q. 14, use the realization of S_3 in the form of the set $\{e, x, y, y^2, xy, xy^2\}$ to explicitly compute all possible one-dimensional complex representations of this group. (Hint: Consider the implications of the relations $x^2 = y^3 = e$, $yx = xy^2$ and $y^2x = xy$.)
What are the kernels of these representations?

37. Use the same type of analysis to show that, up to similarity transformations, there is just one irreducible, two-dimensional representation of the group S_3.

38. A group G acts on the left on a set X. Show that

    (a) A representation of G on the vector space Map $(X, \mathbb{C})$ can be obtained by the transformations

    $$(T(g)f)(x) := f(g^{-1}x).$$

    (b) If X is a finite set show that this representation is unitary with respect to the scalar product on Map $(G, \mathbb{C})$

    $$\langle f_1, f_2 \rangle := \sum_{x \in X} f_1^*(x) f_2(x).$$

39. A very special choice for X in the above is $X = G$ which gives rise to the 'left regular representation' on Map $(G, \mathbb{C})$

    $$(R(g)f)(g') := f(g^{-1}g').$$

    (a) Using the orthonormal basis set $\{f^g \text{ with } g \in G\}$ of Map $(G, \mathbb{C})$ defined by

    $$f^g(a) := \begin{array}{l} 1 \text{ if } g = a \\ 0 \text{ otherwise,} \end{array}$$

    construct a matrix version of the regular representation.

    (b) What are these matrices in the special case of $G = \mathbb{Z}_3$?

    (c) The 'right' regular representation is also defined on Map $(G, \mathbb{C})$, but now with the transformations

$$(S(g)f)(g') := f(g'\,g).$$

Show that this is indeed a representation of G and that, when G is a finite group, it is unitarily equivalent to the left regular representation.

40. Construct the characters of the representations of S_3 found in Q. 36 and 37 and show that they satisfy the orthogonality relations discussed in the lectures.

SOLUTIONS

1. The symmetric group S_N is defined to be the group of bijections/permutations of the set $\{1, 2, \ldots, N\}$ of N objects. Under such a map $\varphi : \{1, 2, \ldots, N\} \to \{1, 2, \ldots, N\}$, the first object 1 can be mapped into any of the N objects. But then the second object 2 can only be mapped into $N-1$ different objects since φ being injective (i.e., one-to-one) implies that $\varphi(1) \neq \varphi(2)$. Similarly, the object 3 can only be mapped by φ into $N-2$ objects, and so on. Thus the total number of different choices for the permutation φ is
$(N-1)(N-2) \ldots 3 \times 2 \times 1 = N!$

2. In order to prove that $G_1 \times G_2$ is a group, we must show that the given product law is associative, that there exists a unit element, and that each element has an inverse.

   (a) To show associativity we note that

   $$((g_1, g_2)(g_1', g_2'))(g_1'', g_2'') = (g_1 g_1', g_2 g_2')(g_1'', g_2'')$$

   $$= ((g_1 g_1')g_1'', (g_2 g_2')g_2'')$$

   $$= (g_1(g_1' g_1''), g_2(g_2' g_2''))$$
   by associativity of G_1 and G_2

   $$= (g_1, g_2)((g_1', g_2')(g_1'', g_2'')).$$

(b) The unit element is clearly (e_1, e_2) where e_1 and e_2 are the unit elements in G_1 and G_2, respectively. Indeed, $(g_1, g_2)(e_1, e_2) = (g_1 e_1, g_2 e_2) = (g_1, g_2)$.

(c) The inverse of (g_1, g_2) is (g_1^{-1}, g_2^{-1}) since

$$(g_1, g_2)(g_1^{-1}, g_2^{-1}) = (g_1 g_1^{-1}, g_2 g_2^{-1}) = (e_1, e_2).$$

Hence $G_1 \times G_2$ is a group.

The number of elements in $G_1 \times G_2$ is clearly the number of pairs of the form (g_1, g_2) that can be formed with $g_1 \in G_1$ and $g_2 \in G_2$. But this is the number of elements in G_1 multiplied by the number of elements in G_2, i.e., $|G_1 \times G_2| = |G_1| \times |G_2|$.

3. Suppose that a subgroup H of $\mathbb{Z}_4 = \{e, a, a^2, a^3\}$ contains the element a. Then, since H is closed under multiplication of its elements, it must contain $a^2 = a \cdot a$ and therefore also $a^3 = (a^2)a$. Thus if H is to be a proper subgroup, it cannot contain a.

Similarly, if H contains a^3 then it contains $(a^3)^2 = a^6 = a^4 \cdot a^2 = e \cdot a^2 = a^2$. Then it must also contain $a^3 \cdot a^2 = a^5 = a^4 \cdot a = e \cdot a = a$; hence if H is to be a proper subgroup, it cannot contain a^3. Thus the only proper subgroup of $\mathbb{Z}_4$ is $\mathbb{Z}_2 := \{e, a^2\}$.

4. In order to show that $j(f)$ is an isomorphism from $\mathrm{Perm}(Y)$ onto a subgroup of $\mathrm{Perm}(X)$ it is necessary to show that

(a) $j(f)$ actually belongs to the subset $\mathrm{Perm}(X)$ of $\mathrm{Map}\,(X, X)$ when f is a permutation of Y.

(b) $j(f_1 \circ f_2) = j(f_1) \circ j(f_2)$ for all $f_1, f_2 \in \mathrm{Perm}(Y)$; i.e., the map j preserves the group law.

(c) The map $f \to j(f)$ is a one-to-one map from $\mathrm{Perm}(X)$ into $\mathrm{Perm}(Y)$.

We will prove these statements in turn.

(a) $j(f) : X \to X$ is defined by $(j(f))(x) := f(x)$ if $x \in Y \subset X$
$$:= x \text{ otherwise.}$$

Thus $j(f)$ maps the points inside the subset Y of X into the same points in Y that the original map f did, whereas if a point $x \in X$ lies outside the subset Y, it is left alone. It is clear that $j(f)$ is a one-to-one map of X into itself and it is also bijective since it has the inverse

$$(j(f))^{-1}(x) := f^{-1}(x) \text{ if } x \in Y \subset X$$

$$:= x \qquad \text{otherwise.}$$

Hence $j(f)$ belongs to Perm(X) for each $f \in$ Perm(Y).

(b) We have

$$(j(f_1 \circ f_2))(x) = (f_1 \circ f_2)(x) = f_1(f_2(x)) \text{ if } x \in Y \subset X$$

$$= x \quad \text{otherwise.}$$

On the other hand, we have

$$(j(f_1) \circ j(f_2))(x) = j(f_1)(j(f_2)(x))$$

$$= f_1(j(f_2)(x)) \text{ if } j(f_2)(x) \in Y \subset X$$

$$= j(f_2)(x) \quad \text{otherwise.}$$

But $j(f_2)(x)$ belongs to the subset Y if and only if x belongs to Y. And in that case $j(f_2)(x) = f_2(x)$, otherwise $j(f_2)(x) = x$. Thus

$$(j(f_1) \circ j(f_2))(x) = f_1(f_2(x)) \text{ if } x \in Y \subset X$$

$$= x \qquad \text{otherwise.}$$

Thus, for all $x \in X$, $(j(f_1) \circ j(f_2))\,(x) = f_1 \circ f_2\,(x)$.

(c) To show that $f \rightarrow j(f)$ is one-to-one suppose that $j(f_1) = j(f_2)$ for some pair of maps $f_1, f_2 \in \mathrm{Perm}(Y)$. Then $f_1(x) = f_2(x)$ for all $x \in Y \subset X$. But this implies that $f_1 = f_2$.

QED.

5. The group $U(n)$ is defined to be the subset of all invertible, $n \times n$ square complex matrices in $GL(n,\ \mathbb{C})$ satisfying the condition $UU^\dagger = \mathbb{1}$. Then,

(a) The diagonal elements of the complex matrix equation $UU^\dagger = \mathbb{1}$ produce a set of n real polynomial constraints on the $2n^2$ real matrix elements of U regarded as an element of $GL(n,\ \mathbb{C})$.

(b) There are a further $2 \times (n^2 - n)/2 = n^2 - n$ real equations coming from the matrix elements lying above the diagonal.

(c) The matrix elements of the equation $UU^\dagger = \mathbb{1}$ coming from below the diagonal do not yield any further constraints as they are merely the complex conjugates of those coming from above the diagonal.

Thus the total number of real variables needed to parametrize the matrices in $U(n)$ is $2n^2 - [n + (n^2 - n)] = n^2$.
As far as the group $SU(n)$ is concerned, the extra condition $\det U = 1$ might appear at first to impose an extra two real constraints. However, the condition $UU^\dagger = \mathbb{1}$ implies that $|\det U| = 1$ and hence the magnitude of $\det U$ is already fixed. Thus the extra condition for $U \in SU(n)$ restricts only the phase of $\det U$, and the number of real variables needed to parametrize matrices in $SU(n)$ is $n^2 - 1$.

6. If $\mu : G_1 \rightarrow G_2$ is a homomorphism with e_1 the unit element in G_1 then we have $\mu(e_1)\,\mu(e_1) = \mu(e_1 \cdot e_1) = \mu(e_1)$ and hence $\mu(e_1)$ is the unit element e_2 in G_2.

Also, for any $g \in G_1$, $\mu(g)\,\mu(g^{-1}) = \mu(gg^{-1}) = \mu(e_1) = e_2$ and hence, by the uniqueness of the inverse of a group element, we have $(\mu(g))^{-1} = \mu(g^{-1})$.

7. Let $G_{x_0} := \{g \in G \mid gx_0 = x_0\}$. If $g_1, g_2 \in G_{x_0}$ we have $(g_1\,g_2)\,x_0 = g_1\,(g_2 x_0) = g_1 x_0 = x_0$ and hence $g_1 g_2 \in G_{x_0}$.

   Similarly, if $gx_0 = x_0$ then $g^{-1}x_0 = g^{-1}(gx_0) = (g^{-1}g)x_0 = ex_0 = x_0$ and hence the subset $G_{x_0} \subset G$ is also closed under inversion. Thus G_{x_0} is a subgroup of G.

8. We have a homomorphism $\mu : G_1 \to G_2$ with Ker $\mu := \{g \in G_1 \mid \mu(g) = e_2\}$. Then if $g_1, g_2 \in$ Ker μ, we have $\mu(g_1 g_2) = \mu(g_1)\,\mu(g_2) = e_2 e_2 = e_2$, and hence the product $g_1 g_2$ also belongs to the kernel of μ.

   Similarly, if $g \in$ Ker μ then $\mu(g^{-1}) = (\mu(g))^{-1}$ [by (6) above] $= e_2^{-1} = e_2$, and hence the kernel of μ is closed under inversion. Thus Ker μ is a subgroup of G_1.

   Now suppose that $g_1, g_2 \in$ Im $\mu \subset G_2$. Then there exists some g_1' and g_2' in G_1 such that $g_1 = \mu(g_1')$ and $g_2 = \mu(g_2')$. Then $g_1 g_2 = \mu(g_1')\,\mu(g_2') = \mu(g_1'\,g_2')$ and hence the product $g_1 g_2$ belongs to Im μ.

   Also, if $g \in$ Im μ then $g = \mu(g')$ for some $g' \in G_1$ and $g^{-1} = (\mu(g'))^{-1} = \mu(g'^{-1})$ and hence the image of μ is closed under inversion too. Thus Im μ is a subgroup of G_2.

9. Let $A = \begin{pmatrix} a & b \\ c & d \end{pmatrix}$ and $A' = \begin{pmatrix} a' & b' \\ c' & d' \end{pmatrix}$ be two matrices in SL(2, $\mathbb{C}$) with corresponding Möbius transformations $z \to \dfrac{az + b}{cz + d}$ and $z \to \dfrac{a'z + b'}{c'z + d'}$. Then the composite of these transformations is

$$z \rightarrow \frac{a'\left(\dfrac{az + b}{cz + d}\right) + b'}{c'\left(\dfrac{az + b}{cz + d}\right) + d'} = \frac{(a'a + b'c)z + (a'b + b'd)}{(c'a + d'c)z + (c'b + d'd)}$$

which is precisely the Möbius transformation induced by the product matrix

$$A'A = \begin{pmatrix} a' & b' \\ c' & d' \end{pmatrix}\begin{pmatrix} a & b \\ c & d \end{pmatrix} = \begin{pmatrix} a'a+b'c & a'b+b'd \\ c'a+d'c & c'b+d'd \end{pmatrix}.$$

Hence the map μ preserves the product structure.

Similarly, it is easy to see that the inverse Möbius transformation to $z \rightarrow \dfrac{az+b}{cz+d}$ is generated by the inverse matrix of $\begin{pmatrix} a & b \\ c & d \end{pmatrix}$.

Thus the map μ is a homomorphism.

10. (a) (i) If $ab = cb$, then multiplying both sides by b^{-1} gives $(ab)b^{-1} = (cb)b^{-1}$. But then, by associativity, $a(bb^{-1}) = c(bb^{-1})$ and so $ae = ce$, and hence $a = c$.
    (ii) Similarly for $ba = bc$.

   (b) $ab(b^{-1}a^{-1}) = a((bb^{-1})a^{-1})$ (by associativity)
   $= a(ea^{-1}) = aa^{-1} = e$.
   But inverses in a group are unique, and hence $(ab)^{-1} = b^{-1}a^{-1}$.

11. In the space $\text{Map}(X, G)$, we have defined $(f_1 f_2)(x) := f_1(x) f_2(x)$. In order to show that this gives a group structure for $\text{Map}(X, G)$, we must
   (a) show that this combination law is associative,
   (b) find a suitable unit element,
   (c) find an inverse for each f in $\text{Map}(X, G)$.

(a) We have,

$$[(f_1 f_2) f_3](x) = (f_1 f_2)(x) f_3(x) = (f_1(x) f_2(x)) f_3(x)$$
$$= f_1(x) (f_2(x) f_3(x)) \quad \text{(by associativity of } G\text{)}$$
$$= f_1(x) (f_2 f_3)(x) = [f_1 (f_2 f_3)](x)$$

and, since this is true for all x in X, it implies that $(f_1 f_2) f_3 = f_1(f_2 f_3)$, i.e., the combination law in Map(X, G) is associative.

(b) A suitable unit element in Map(X, G) is the constant map e defined as $e(x) = e_G$ (the unit element in G) for all x in X. This is indeed a unit since, for any f in Map(X, G),

$$(ef)(x) = e(x) f(x) = e_G f(x) = f(x)$$

and similarly for fe.

(c) An inverse for f in Map(X, G) is the function f^{-1} defined by $f^{-1}(x) := (f(x))^{-1}$ (the inverse in G).
Since

$$(ff^{-1})(x) = f(x) f^{-1}(x)$$
$$= f(x) (f(x))^{-1}$$
$$= e_G$$
$$= e(x)$$

and so, since this is true for all x in X, $ff^{-1} = e$ in Map(X, G). Similarly, $f^{-1} f = e$ in Map(X, G).

QED.

12. The Möbius transformations of the complex plane are of the form

$$z \to \frac{az + b}{cz + d} \tag{1}$$

with

$$ad - bc = 1 . \tag{2}$$

To show that they form a subgroup of the group of all bijections of the complex plane onto itself we must prove that:
(a) the identity transformation is a Möbius transformation,
(b) the product of two Möbius transformations is Möbius,
(c) the inverse of a Möbius transformation is Möbius.

(a) The identity transformation is simply $z \to z$ which is a Möbius transformation with $a = d = 1$ and $c = b = 0$ (and then $ad - bc = 1$ too).

(b) Consider two Möbius transformations

$$z \to \frac{az + b}{cz + d} \quad \text{and} \quad z \to \frac{a'z + b'}{c'z + d'}$$

and let us perform the first one followed by the second. This gives the transformation

$$z \to \frac{a'\left(\frac{az+b}{cz+d}\right) + b'}{c'\left(\frac{az+b}{cz+d}\right) + d'} = \frac{a'(az+b) + b'(cz+d)}{c'(az+b) + d'(cz+d)}$$

$$= \frac{(a'a + b'c)z + (a'b + b'd)}{(c'a + d'c)z + (c'b + d'd)} \tag{3}$$

which is of the right form (1). Now we must check (2).

$$(a'a + b'c)(c'b + d'd) - (a'b + b'd)(c'a + d'c)$$

$$= a'c'(ab - ba) + a'd'(ad - bc)$$
$$\quad + b'c'(cb - da) + b'd'(cd - dc)$$
$$= a'd' - b'c' = 1 .$$

Thus the product of two Möbius transformations is Möbius.

(c) To see if the inverse of a Möbius transformation is Möbius, we could try and find a', b', c', d' in Eq. (3) such that

$$(a'a + b'c) = 1 = (c'b + d'd);$$
$$(a'b + b'd) = 0 = (c'a + d'c)$$

as the Möbius transformation parametrized by a', b', c', d' would then be the inverse of the one parametrized by a, b, c, d.

Elementary algebra shows that these equations can indeed be solved with the solution (taking into account $ad - bc = 1$)

$$a' = d, \, b' = -b, \, c' = -c, \, d' = a$$

and hence the inverse of a Möbius transformation is itself a Möbius transformation.

QED.

13. Let the group elements in the group of order 4 be $\{e, a, b, c\}$. Since these are distinct elements we must have $ab \neq a$ and $ab \neq b$ and hence there are just two possibilities for the product ab, i.e., either $ab = e$ or $ab = c$. Let us consider these two cases in turn.

Case (a) $ab = e$

$b(ab) = be = b$, and so $(ba)b = b(ab) = b$. Thus $ba = e$. (1)
$ac = e$ or b. But $ac = e$ would imply that $bac = b$ and hence [from (1)] that $c = b$. But $c \neq b$ and hence we deduce that
$$ac = b. \tag{2}$$

Similarly, $ca = e$ or b. But $ca = e$ implies that $cab = b$ and hence that $c = b$ (since we are assuming that $ab = e$). Thus

$$ca = b. \tag{3}$$

$bc = e$ or a. But $bc = e \Rightarrow abc = a \Rightarrow c = a$. Thus $bc = a$. (4)

$cb = e$ or a. But $cb = e \Rightarrow cba = a \Rightarrow c = a$ [from (1)]. Thus

$$cb = a. \tag{5}$$

$a^2 = e$ implies $a^2 = ab \Rightarrow a = b$. Therefore $a^2 \neq e$.

$a^2 = b$ implies $a^2 = ac$ [from (2)] $\Rightarrow a = c$. Thus $a^2 \neq b$

$$\Rightarrow a^2 = c. \tag{6}$$

$b^2 = acb$ [from (2)] $= a^2$ [from (5)]. Thus (6) $\Rightarrow b^2 = c$. (7)

$c^2 = a^2b^2$ [from (6–7)] $= a(ab)b = aeb = ab = e$. Thus $c^2 = e$.

Thus the group table is

	e	a	b	c
e	e	a	b	c
a	a	c	e	b
b	b	e	c	a
c	c	b	a	e

Note that $c = b^2$ and $a = bc = b^3$ and hence we can identify this group with the cyclic group

$$\mathbb{Z}_4 = \{e, b, b^2, b^3\}.$$

Case (b) $ab = c$

If we have any of ba, bc, cb, ac or ca equal to e, then we will merely recover case (a) with some of the letters exchanged. Thus

$$ab = ba = c \tag{8}$$
$$bc = cb = a \tag{9}$$
$$ac = ca = b. \tag{10}$$

But (8) $\Rightarrow a^2b = ac = b$ [from (10)] and so $a^2 = e$.

But (9) $\Rightarrow b^2c = ba = c$ [from (8)] and so $b^2 = e$.

But (10) $\Rightarrow c^2a = cb = a$ [from (9)] and so $c^2 = e$.

Thus the group table is

	e	a	b	c
e	e	a	b	c
a	a	e	c	b
b	b	c	e	a
c	c	b	a	e

Note that this is precisely the group table of $V_4 \cong \mathbb{Z}_2 \times \mathbb{Z}_2$.

QED.

14. I will introduce the schematic notation abc for the permutation $\begin{pmatrix} 1 \to a \\ 2 \to b \\ 3 \to c \end{pmatrix}$ thinking of S_3 as Perm $\{1, 2, 3\}$.

In constructing the group table, it must be remembered that the conventions I am following in relation to the rows and columns gives the generic form

	e	g_1	g_2	g_3	$\cdots$
e	e	g_1	g_2	g_3	$\cdots$
g_1	g_1	$g_1 g_1$	$g_1 g_2$	$g_1 g_3$	$\cdots$
g_2	g_2	$g_2 g_1$	$g_2 g_2$	$g_2 g_3$	$\cdots$
g_3	g_3	$g_3 g_1$	$g_3 g_2$	$g_3 g_3$	$\cdots$
$\vdots$	$\vdots$	$\vdots$	$\vdots$	$\vdots$	$\vdots$

This is important to get right since for a non-abelian group there will be g_i and g_j such that $g_i g_j \neq g_j g_i$ and then one must get the row and column assignments correct.

Remembering that, for a permutation group of mappings, the product law $g_1 g_2$ means that $g_1 g_2$ is obtained by doing the map g_2 first and then the map g_1, the group table for S_3 can be constructed immediately in 'ones head' term by term as follows

	123	213	132	321	312	231
123	123	213	132	321	312	231
213	213	123	231	312	321	132
132	132	312	123	231	213	321
321	321	231	312	123	132	213
312	312	132	321	213	231	123
231	231	321	213	132	123	312

The group can be represented in the way requested by inspecting this table and making the assignments

$$e := 123, x := 213, y := 312, y^2 = 231, xy = 321, xy^2 = 132$$

which are consistent [eg. it is true that $(312)^2 = 231$] and which satisfy the relations

$$x^2 = y^3 = e, yx = xy^2 \quad \text{and} \quad y^2 x = xy.$$

The group is clearly non-abelian since, for example,

$$132 \ 213 = 312 \quad \text{whereas } 213 \ 132 = 231 \neq 312.$$

$$\text{QED.}$$

15. Let $T := \left\{ M \text{ in } M(2, \mathbb{C}) \text{ such that } M = \begin{pmatrix} a & c \\ 0 & b \end{pmatrix} \text{ with } ab \neq 0 \right\}.$

Then the first step is to show that T is a subset of GL(2, $\mathbb{C}$). But, if M is in T then

$$\text{Det}(M) = \text{Det} \begin{pmatrix} a & c \\ 0 & b \end{pmatrix} = ab \neq 0$$

and hence M does lie in GL(2, $\mathbb{C}$).

Now: (a) The unit matrix is the unit element in GL(2, $\mathbb{C}$) and it clearly lies also in the subset T.

(b) Let $M = \begin{pmatrix} a & c \\ 0 & b \end{pmatrix}$ and $M' = \begin{pmatrix} a' & c' \\ 0 & b' \end{pmatrix}$ belong to T.

Then, $MM' = \begin{pmatrix} aa' & ac' + cb' \\ 0 & bb' \end{pmatrix}$ which is of the correct triangular form, and with $(aa')(bb') \neq 0$. Thus MM' belongs to T.

(c) An inverse for M lying in T will exist if the expression above can be solved for a', b', c' in the equation

$$\begin{pmatrix} aa' & ac' + cb' \\ 0 & bb' \end{pmatrix} = \begin{pmatrix} 1 & 0 \\ 0 & 1 \end{pmatrix}. \quad \text{But this gives}$$

$$a' = 1/a$$
$$b' = 1/b$$
$$c' = -c/ab.$$

Thus the inverse of $\begin{pmatrix} a & c \\ 0 & b \end{pmatrix}$ is $\begin{pmatrix} \frac{1}{a} & \frac{-c}{ab} \\ 0 & \frac{1}{b} \end{pmatrix}$ which belongs to T.

Thus the three conditions specifying a subgroup are satisfied.

QED.

16. An automorphism of a group is a map $\mu : G \rightarrow G$ that is an isomorphism of G onto itself. The natural product law to place on the set $\text{Aut}(G)$ of all automorphisms is the usual composition of mappings. Thus, if μ_1, μ_2 are in $\text{Aut}(G)$, we define

$$(\mu_1 \mu_2)(g) := \mu_1 \circ \mu_2(g) = \mu_1(\mu_2(g)) \qquad (1)$$

and the first step is to show that $\mu_1 \mu_2$ thus defined is itself an automorphism of G. The composition of two isomorphisms is of course an isomorphism and so all that is necessary is to show that $\mu_1 \mu_2$ preserves the group law

$$\mu(gg') = \mu(g)\,\mu(g') \text{ for all } g, g' \text{ in } G. \tag{2}$$

Thus,

$$(\mu_1\mu_2)(gg') = \mu_1[\mu_2(gg')] = \mu_1[\mu_2(g)\,\mu_2(g')]$$
$$[\text{since } \mu_2 \in \text{Aut}(G)]$$
$$= \mu_1(\mu_2(g))\,\mu_1(\mu_2(g')) \text{ [since } \mu_1 \text{ is in Aut}(G)]$$
$$= (\mu_1\mu_2)(g)\,(\mu_1\mu_2)(g') \quad \text{from (1)}$$

and hence $\mu_1\mu_2$ belongs to Aut(G).

Now we must show that Aut(G) is a group. This involves proving (a) associativity; (b) the existence of a unit; (c) the existence of an inverse for each μ in Aut(G).

(a) If μ_1, μ_2, μ_3 belong to Aut(G) then

$$\begin{aligned}[\mu_1(\mu_2\mu_3)](g) &= \mu_1[(\mu_2\mu_3)(g)] \\ &= \mu_1[\mu_2(\mu_3(g))] \\ &= (\mu_1\mu_2)(\mu_3(g)) \\ &= [(\mu_1\mu_2)\mu_3](g)\end{aligned}$$

which, since this is true for all g in G, implies associativity. [**Note.** This is just a repeat of the proof that, in general, composition of maps in Map(X, X) is associative.]

(b) The identity map in the composition of mappings is $id_G : G \to G$ with $id_G(g) := g$ for all g in G. This is clearly an automorphism and so it will also serve as an identity in the group Aut(G).

(c) Since every μ in Aut(G) is a bijection, it has an inverse $\mu^{-1} : G \to G$ in the general group Perm(G) defined by $\mu(\mu^{-1}(g)) := g$ and the only task is to show that μ^{-1} thus defined is itself an automorphism. Thus we study

$$\mu(\mu^{-1}(gg')) = gg' \text{ (by definition).} \qquad (3)$$

On the other hand,

$$\mu[\mu^{-1}(g)\,\mu^{-1}(g')] = \mu[\mu^{-1}(g)]\,\mu[\mu^{-1}(g')] \qquad (4)$$

since μ is an isomorphism. But the right hand side of (4) $= gg'$ and hence, on comparing (4) with (3) we see that

$$\mu(\mu^{-1}(gg')) = \mu[\mu^{-1}(g)\mu^{-1}(g')]$$

and hence, since μ is injective, $\mu^{-1}(gg') = \mu^{-1}(g)\,\mu^{-1}(g')$. Thus μ^{-1} is an automorphism of G, and we have completed the proof that $\text{Aut}(G)$ is a group.

$G = \mathbb{Z}_2$

Let us write $\mathbb{Z}_2 = \{e, a\}$ with $a^2 = e$, and consider an automorphism μ. This can only map a into e or a itself. But, since $\mu(e) = e$ always for any homomorphism, $\mu(a) = e$ would imply that μ was not one-to-one, and hence it could not belong to $\text{Aut}(\mathbb{Z}_2)$. Hence we must have $\mu(a) = a$. But this just defines the identity automorphism (which of course always exists) and so we conclude that $\text{Aut}(\mathbb{Z}_2) \cong \{id_{\mathbb{Z}_2}\}$, i.e., the trivial group with just one element.

$G = \mathbb{Z}_4$

Write $\mathbb{Z}_4 = \{e, a, a^2, a^3\}$ with $a^4 = e$. For any cyclic group $\mathbb{Z}_n = \{e, a, \ldots, a^{n-1}\}$ with $a^n = e$, the effect of an automorphism is uniquely determined by its value on a since $\mu(a^m) = (\mu(a))^m$ by iterating (1) m-times. Thus it suffices to look at the four possible values for $\mu(a)$ and reject those that do not lead to a bijective map [but, by the remarks just made, *any* choice for $\mu(a)$ will produce a homomorphism from G into itself. Maps of this more general type are known as *endomorphisms* of G; so an automorphism is a bijective endomorphism]

(i) $\mu(a) = e$ does not produce a bijection since we have $\mu(e) = e$,

(ii) $\mu(a) = a$ implies $\mu(a^q) = a^q$ and so this just produces $id_{\mathbb{Z}_4}$,

(iii) $\mu(a) = a^2$ implies $\mu(a^2) = a^4 = e$, and so this is not injective either,

(iv) finally, $\mu(a) = a^3$ implies $\mu(a^2) = a^6 = a^2$ and $\mu(a^3) = a^9 = a$, and this *does* give a bijection. We conclude that $\text{Aut}(\mathbb{Z}_4)$ has two elements and must therefore be the group $\mathbb{Z}_2$. This is confirmed by computing $\mu^2(a) = \mu(\mu(a)) = \mu(a^3) = a$ so that $\mu^2 = id_{\mathbb{Z}_2}$ in $\text{Aut}(\mathbb{Z}_2)$.

QED.

[**Note.** To save space, from now on the symbol $x \in X$ will be used rather than saying each time that 'x is a member of X'.]

17. (a) Since both H and K are subgroups of G, it follows that $e \in H$ and $e \in K$. Thus e belongs to their intersection $H \cap K$.

Let h and h' belong to $H \cap K$. Then $h, h' \in H$ and $h, h' \in K$. But H and K are subgroups of G and hence $hh' \in H$ and $hh' \in K$. Thus $hh' \in H \cap K$.

Let $h \in H \cap K$. Then $h \in H$ and $h \in K$ and therefore, since H and K are subgroups, $h^{-1} \in H$ and $h^{-1} \in K$. Thus $h^{-1} \in H \cap K$.

Hence $H \cap K$ satisfies the three requirements for a subset of a group G to be a subgroup.

(b) The obvious thing to do is to try and find two suitable $\mathbb{Z}_2$ subgroups of S_3 since these are the simplest non-trivial groups.

Looking at the group table in Q. 14, we can see that two such subgroups are

$$H := \{123, 213\} \quad \text{and} \quad K := \{123, 132\}.$$

These are certainly subgroups and their union is

$$H \cup K = \{123, 213, 132\}.$$

But this is not a subgroup of S_3 since $(213)(132) = (231)$ and this is *not* an element of $H \cup K$.

(c) If $HK = KH$ then any $kh \in KH$ can be written in the form $h'k'$ for some $h' \in H$ and $k' \in K$. In particular, if $h_1, h_2 \in H$ and $k_1, k_2 \in K$ then there exists h', k' such that $k_1 h_2 = h' k'$. Then,

$$h_1 k_1 h_2 k_2 = h_1 h' k' k_2, \text{ i.e., } (h_1 k_1)(h_2 k_2) \in HK.$$

Similarly, if $hk \in HK$ then $(hk)^{-1} = k^{-1} h^{-1} = h'k'$ for some $h' \in H$ and $k' \in K$. Thus $hk \in HK$ implies that $(hk)^{-1} \in HK$.

Clearly $e \in HK$ since $e \in H$ and $e \in K$. Thus HK satisfies the three conditions for being a subgroup.

Conversely, if HK is a subgroup of G then every element of HK is an inverse of a unique element of HK. Thus

$$HK = \{(hk)^{-1} \text{ with } h \in H \text{ and } k \in K\}$$
$$= \{k^{-1} h^{-1} \text{ with } h \in H \text{ and } k \in K\} = KH.$$

$$\text{QED.}$$

(d) If H is a subgroup of G then $b \in H$ implies $b^{-1} \in H$ and therefore $a \in H$ and $b \in H$ implies $ab^{-1} \in H$.

Conversely, if $a, b \in H$ implies ab^{-1} in H then, since H is non-empty, there exists at least one $g \in H$ and hence

$e = gg^{-1} \in H$. But then, if h is any member of H, e, $h \in H$ and therefore $eh^{-1} \in H$ and so $h^{-1} \in H$.

Finally, if h_1, $h_2 \in H$ then h_1, $h_2^{-1} \in H$ and therefore $h_1 h_2 \in H$.

Thus H is a subgroup of G.

18. (a) If $H_2 = gH_1g^{-1}$, we define a map $i : H_1 \to H_2$ by $i(h) := ghg^{-1}$. Then,

    (i) i is one-to-one since $i(h_1) = i(h_2) \Rightarrow gh_1g^{-1} = gh_2g^{-1} \Rightarrow h_1 = h_2$.

    (ii) i is surjective since $H_2 = gH_1g^{-1}$.

    (iii) $i(h_1 h_2) = gh_1 h_2 g^{-1} = (gh_1 g^{-1})(gh_2 g^{-1}) = i(h_1)\, i(h_2)$.

    Thus $i : H_1 \to H_2$ is an isomorphism.

(b) One $\mathbb{Z}_2$ subgroup of S_3 is $H := \{123, 213\}$. We can find a conjugate subgroup by choosing any $g \in S_3$ such that $g(213)g^{-1} \neq 213$.

Using the group table in Q. 14, we see that, for example, $(312)^{-1} = 231$ and $(312)(213)(231) = (312)(132) = 321 \neq 213$. Thus a pair of conjugate subgroups in S_3 is $H_1 := \{123, 213\}$ and $H_2 := \{123, 321\}$. Both are isomorphic to $\mathbb{Z}_2$.

19. Since $l_g : X \to X$ is a left action of G on X we have the relation

$$l_{g_2} \circ l_{g_1} = l_{g_2 g_1} \text{ for all } g_1, g_2 \text{ in } G.$$

Then, if $r_g := l_{g^{-1}}$, we have

$$r_{g_2} \circ r_{g_1} = l_{g_2^{-1}} \circ l_{g_1^{-1}} = l_{g_2^{-1} g_1^{-1}} = l_{(g_1 g_2)^{-1}} = r_{g_1 g_2}.$$

Thus $r_g : X \to X$ is a right action.

20. (a) The orbits through the points x_1 and x_2 in X are

$$O_{x_1} = \{gx_1 \text{ with } g \in G\}, \quad O_{x_2} = \{gx_2 \text{ with } g \in G\}.$$

Let $y \in O_{x_1} \cap O_{x_2}$. Then $y = g_1 x_1 = g_2 x_2$ for some g_1, $g_2 \in G$. Then if gx_1 is any other point on the orbit O_{x_1}, we have $gx_1 = gg_1^{-1}g_2 x_2$ and so gx_1 lies also on the orbit O_{x_2}. Thus $O_{x_1} \subset O_{x_2}$. Similarly, $O_{x_2} \subset O_{x_1}$ and therefore, if $O_{x_1} \cap O_{x_2}$ is non-empty, we must have $O_{x_1} = O_{x_2}$.

(b)

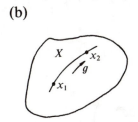

Since x_1 and x_2 lie on the same orbit of the G-action, there exists some $g \in G$ such that $gx_1 = x_2 (\Rightarrow x_1 = g^{-1}x_2)$. Let $h \in G_{x_1}$, i.e., $hx_1 = x_1$. Then
$$(ghg^{-1})x_2 = gh(g^{-1}x_2)$$
$$= (gh)x_1 = g(hx_1) = gx_1 = x_2.$$

Thus we see that $ghg^{-1} \in G_{x_2}$ and so $gG_{x_1}g^{-1} \subset G_{x_2}$. Similarly, $g^{-1}G_{x_2}g \subset G_{x_1}$ and hence $gG_{x_1}g^{-1} = G_{x_2}$.

QED.

21. First, we must show that the centre $C(G)$ of G is a subgroup of G. So, let $a, b \in C(G)$. Then, for all $g \in G$,

$$(ab)g(ab)^{-1} = a(bgb^{-1})a^{-1} = aga^{-1} = g$$

and hence $a, b \in C(G)$ implies that $ab \in C(G)$. The unit element e clearly belongs to $C(G)$, and if $a \in C(G)$ then $aga^{-1} = g \Rightarrow a^{-1}ga = g$, i.e., $a^{-1} \in C(G)$. Thus $C(G)$ is a subgroup of G.

To show that it is a normal subgroup note that if $a \in C(G)$ then, for all $g \in G$, $gag^{-1} = gg^{-1}a = a$ so that $C(G)$ is a normal subgroup.

22. From Eq. (1.3.20) of the lecture notes, we know that the most general element of SU2 is of the form

$$\begin{pmatrix} a & b \\ -b^* & a^* \end{pmatrix} \quad \text{with } |a|^2 + |b|^2 = 1. \tag{1}$$

Let $\begin{pmatrix} c & d \\ -d^* & c^* \end{pmatrix}$ be an element of $C(\text{SU2})$. Then for all matrices of the form (1), we must have $\begin{pmatrix} a & b \\ -b^* & a^* \end{pmatrix}\begin{pmatrix} c & d \\ -d^* & c^* \end{pmatrix}$

$$= \begin{pmatrix} c & d \\ -d^* & c^* \end{pmatrix}\begin{pmatrix} a & b \\ -b^* & a^* \end{pmatrix}.$$

Therefore,

$$\begin{pmatrix} ac - bd^* & ad + bc^* \\ -b^*c - a^*d^* & -b^*d + a^*c^* \end{pmatrix}$$

$$= \begin{pmatrix} ca - db^* & cb + da^* \\ -d^*a - c^*b^* & -d^*b + c^*a^* \end{pmatrix}. \tag{2}$$

Equating the 1, 1 matrix elements gives $bd^* = b^*d$, which has to be true for all b such that $|a|^2 + |b|^2 = 1$. In particular, it must be satisfied for $b = 1$, which gives $d = d^*$, and $b = \sqrt{-1}$, which gives $-d = d^*$. Hence, we msut have $d = 0$. The remaining matrix elements of Eq. (2) imply that $c = c^*$, so that c is real. But we also have $|c|^2 + |d|^2 = 1$ and therefore $c = \pm 1$. Hence, the centre of SU2 consists of the pair of matrices

$$\begin{pmatrix} 1 & 0 \\ 0 & 1 \end{pmatrix} \text{ and } \begin{pmatrix} -1 & 0 \\ 0 & -1 \end{pmatrix} \quad \text{which form a } \mathbb{Z}_2 \text{ subgroup of SU2.}$$

23. First, we must show that the set D of matrices $\begin{pmatrix} 1 & d \\ 0 & 1 \end{pmatrix}$ is a subgroup of the group $T := \left\{ \begin{pmatrix} a & c \\ 0 & b \end{pmatrix} \text{ with } ab \neq 0 \right\}$. It is clear that $\mathbb{1} \in D$. And $\begin{pmatrix} 1 & d \\ 0 & 1 \end{pmatrix}\begin{pmatrix} 1 & d' \\ 0 & 1 \end{pmatrix} = \begin{pmatrix} 1 & d' + d \\ 0 & 1 \end{pmatrix}$ so that D is closed under multiplication. The inverse of $\begin{pmatrix} 1 & d \\ 0 & 1 \end{pmatrix}$ is $\begin{pmatrix} 1 & -d \\ 0 & 1 \end{pmatrix}$ which also belongs to D. Hence D is a subgroup of T. But $\begin{pmatrix} a & c \\ 0 & b \end{pmatrix}\begin{pmatrix} 1 & d \\ 0 & 1 \end{pmatrix}\begin{pmatrix} 1/a & -c/ab \\ 0 & 1/b \end{pmatrix} = \begin{pmatrix} 1 & ad/b \\ 0 & 1 \end{pmatrix}$ which belongs to D. Hence D is a normal subgroup of T.

QED.

24. The map $\mu : \mathbb{R} \to \mathbb{R}_+$ defined by $\mu(x) := e^x$ satisfies

$$\mu(x+y) = e^{x+y} = e^x e^y = \mu(x)\,\mu(y)$$

and hence μ is a homomorphism. It is a map 'onto' $\mathbb{R}_+$ since every positive real number x can be written as the exponential of a real number, viz. $x = e^{(\log x)}$.
The kernel of this homomorphism is Ker $\mu = \{x \in \mathbb{R}$ such that $e^x = 1\}$, since the number '1' is the unit element in $\mathbb{R}_+$. But the only such x is the number 0. Hence Ker $(\mu) = \{0\}$.
Note. Since μ is an injective homomorphism onto $\mathbb{R}_+$, it is an isomorphism from $\mathbb{R}$ to $\mathbb{R}_+$.

25. We have $U(1) = \{\lambda \in \mathbb{C}$ such that $\lambda\lambda^* = 1\}$.
Motivated by the previous question, we try the map $\mu : \mathbb{R} \to U(1)$ defined by

$$\mu(x) := e^{ix}.$$

As before, we have $\mu(x+y) = e^{i(x+y)} = e^{ix} e^{iy}$ so that μ is a surjective homomorphism from $\mathbb{R}$ onto $U(1)$.

This time, however, $\text{Ker}(\mu) = \{x \in \mathbb{R} \text{ such that } e^{ix} = 1\}$, which is isomorphic to the additive group of the integers $\mathbb{Z}$ since $e^{i2\pi n} = 1$ for all $n \in \mathbb{Z}$.

Furthermore, since $\mathbb{R}$ is an abelian group, the integers $\mathbb{Z}$ form a normal subgroup and hence the space of cosets $\mathbb{R}/\mathbb{Z}$ can be given a group structure.

But then, according to the theorem in Sec. 1.3, the surjective homomorphism $\mu: \mathbb{R} \rightarrow U(1)$ induces an isomorphism $\mathbb{R}/\mathbb{Z} \cong U(1)$.

26. (a) The basic idea behind this question is that a function $f: X \rightarrow \mathbb{C}$ can take on independent values at each of the N points in X, and hence $\text{Map}(X, \mathbb{C})$ is N-dimensional. This suggests that we try the following function, where $\mathbf{x}_1, \mathbf{x}_2, \ldots, \mathbf{x}_N$ is an enumeration of the points in X

$$i : \text{Map}(X, \mathbb{C}) \rightarrow \mathbb{C}^N \qquad (1)$$
$$i(f) := (f(\mathbf{x}_1), f(\mathbf{x}_2), \ldots, f(\mathbf{x}_N)).$$

If $\mu_1, \mu_2 \in \mathbb{C}$ and $f_1, f_2 \in \text{Map}(X, \mathbb{C})$, then

$$i(\mu_1 f_1 + \mu_2 f_2)$$

$$= ((\mu_1 f_1 + \mu_2 f_2)(\mathbf{x}_1), \ldots, (\mu_1 f_1 + \mu_2 f_2)(\mathbf{x}_N))$$

$$= (\mu_1 f_1(\mathbf{x}_1) + \mu_2 f_2(\mathbf{x}_1), \ldots, \mu_1 f_1(\mathbf{x}_N) + \mu_2 f_2(\mathbf{x}_N))$$

$$= \mu_1 i(f_1) + \mu_2 i(f_2)$$

and so $i: \text{Map}(X, \mathbb{C}) \rightarrow \mathbb{C}^N$ is linear.

It is clearly surjective since, given $(a_1, a_2, \ldots, a_N) \in \mathbb{C}^N$, define a function $f \in \text{Map}(X, \mathbb{C})$ by $f(\mathbf{x}_i) := a_i$ for $i = 1, \ldots, N$. Then $i(f) = (a_1, \ldots, a_N)$.

Finally, it is one-to-one since if $i(f_1) = i(f_2)$ for some f_1, f_2 in Map$(X, \mathbb{C})$, then $i(f_1 - f_2) = 0$ which implies that $f_1(\mathbf{x}_i) = f_2(\mathbf{x}_i)$ for all $i = 1, \ldots, N$. And this implies that $f_1 = f_2$ since a function is uniquely determined by its values on all points in X.

Thus the map i is a bijective linear map from the vector space Map$(X, \mathbb{C}) \rightarrow \mathbb{C}^N$, which is precisely the condition that they be isomorphic.

(b) It is clear from (1) that the function f mapped by i into, for example, the basis vector $(1, 0, \ldots, 0)$ is defined by $f(\mathbf{x}_1)$ $:= 1$ and $f(\mathbf{x}_i) := 0$ for $i = 2, \ldots, N$. In general, the vector in Map$(X, \mathbb{C})$ corresponding to the basis vector $(0, 0, \ldots, 1, 0, 0, \ldots, 0)$ (with the '1' in the ith position) is the function $e_i : X \rightarrow \mathbb{C}$ defined by

$$e_i(\mathbf{x}_j) := 1 \text{ if } i = j \qquad i = 1, \ldots, N.$$
$$0 \text{ otherwise.} \tag{2}$$

Evidently, the set of functions $\{e_1, e_2, \ldots, e_N\}$ is a basis set for Map$(X, \mathbb{C})$.

(c) If we define $\langle f_1, f_2 \rangle := \sum_{\mathbf{x} \in X} f_1^*(\mathbf{x}) f_2(\mathbf{x}) \equiv \sum_{i=1}^{N} f_1^*(\mathbf{x}_i) f_2(\mathbf{x}_i)$

$$\tag{3}$$

then it is obvious that $\langle f, (\mu_1 f_1 + \mu_2 f_2) \rangle = \mu_1 \langle f, f_1 \rangle + \mu_2 \langle f, f_2 \rangle$ for all $\mu_1, \mu_2 \in \mathbb{C}$ and all f, f_1, f_2 in Map$(X, \mathbb{C})$. Furthermore,

$$\langle f_1, f_2 \rangle^* = \sum_{i=1}^{N} f_1(\mathbf{x}_i) f_2^*(\mathbf{x}_i) = \langle f_2, f_1 \rangle \text{ as required.}$$

Finally, $\langle f, f \rangle = \sum_{i=1}^{N} |f(\mathbf{x}_i)|^2$ which is ≥ 0 and $= 0$ only if

$f = 0$. Hence, Eq. (3) defines a scalar product on Map($X, \mathbb{C}$).

(d) The usual inner product on $\mathbb{C}^N$ is $\langle a, b \rangle^{\mathbb{C}^N} = \sum_{i=1}^{N} a_i^* b_i$,

where $a = (a_1, a_2, \ldots, a_N)$ and $b = (b_1, b_2, \ldots, b_N)$.

Thus, from Eqs. (1) and (3), we see that

$$\langle f_1, f_2 \rangle = \langle i(f_1), i(f_2) \rangle^{\mathbb{C}^N} \qquad (4)$$

which is to say that the scalar product on Map ($X, \mathbb{C}$) defined by Eq. (3) is equal to the scalar product induced via the isomorphism $i : \text{Map } (X, \mathbb{C}) \rightarrow \mathbb{C}^N$ from the usual scalar product on $\mathbb{C}^N$.

27. The central point here is that an $n \times n$ matrix has precisely n^2 independent entries and hence that these can be regarded in some appropriate sense as the n^2 components of a vector in $\mathbb{C}^{n^2}$. There are several ways in which this correspondence can be set up but they are all based on this basic idea. One concrete example is to define the matrix corresponding to the vector $(a_1, a_2, \ldots, a_{n^2})$ to be

$$\begin{pmatrix} a_1 & a_{n+1} & a_{2n+1} & \cdots & a_{n^2-n+1} \\ a_2 & a_{n+2} & a_{2n+2} & \cdots & a_{n^2-n+2} \\ \vdots & \vdots & \vdots & & \vdots \\ a_n & a_{2n} & a_{3n} & & a_{n^2} \end{pmatrix}$$

i.e., $i : \mathbb{C}^{n^2} \rightarrow M(n, \mathbb{C})$ is defined by

$$(i(a))_{jk} := a_{j+(k-1)n} \quad \text{for} \quad j, k = 1, \ldots, n. \qquad (1)$$

In the case of $n=2$, we have $i(a_1, a_2, a_3, a_4) = \begin{pmatrix} a_1 & a_3 \\ a_2 & a_4 \end{pmatrix}$, and hence the inner product $\langle A, B \rangle^{M(2,\mathbb{C})}$ between the matrices A and B in $M(2, \mathbb{C})$ whose matrix elements are labelled as

$$A = \begin{pmatrix} a_1 & a_3 \\ a_2 & a_4 \end{pmatrix} \text{ and } B = \begin{pmatrix} b_1 & b_3 \\ b_2 & b_4 \end{pmatrix},$$

is induced from that on $\mathbb{C}^4$ to be

$$\langle A, B \rangle^{M(2,\mathbb{C})} = \langle i(a), i(b) \rangle^{M(2,\mathbb{C})} := \langle a, b \rangle^{\mathbb{C}^4} \tag{2}$$

in which $\langle a,b \rangle^{\mathbb{C}^4}$ is the usual inner product $\sum_{i=1}^{4} a_i^* b_i$ on $\mathbb{C}^4$ and we are thinking of the matrices A and B as being the images under the isomorphism $i : \mathbb{C}^4 \rightarrow M(2, \mathbb{C})$ of the vectors (a_1, a_2, a_3, a_4) and (b_1, b_2, b_3, b_4), respectively. Rewriting this in terms of the matrix elements of A and B, we find

$$\langle A, B \rangle^{M(2,\mathbb{C})} := A_{11}^* B_{11} + A_{21}^* B_{21} + A_{12}^* B_{12} + A_{22}^* B_{22}. \tag{3}$$

It seems natural to try to write the right hand side in terms of the product of two matrices, and this can be done if we recall that the adjoint of a matrix is defined by $(A^\dagger)_{jk} := A_{kj}^*$. Using this, we can rewrite Eq. (3) as

$$\langle A, B \rangle^{M(2,\mathbb{C})} := A_{11}^\dagger B_{11} + A_{12}^\dagger B_{21} + A_{21}^\dagger B_{12} + A_{22}^\dagger B_{22}$$

$$= (A^\dagger B)_{11} + (A^\dagger B)_{22}.$$

Thus the final (rather elegant) result is

$$\langle A, B \rangle^{M(2,\mathbb{C})} := \mathrm{Tr}\,(A^\dagger B). \tag{4}$$

It should now be clear how to generalize the method to $M(n, \mathbb{C})$. The result obtained is again Eq. (4).

28. (a) The dimension of a vector space is defined to be the maximum number of linearly independent vectors that it contains. The key point of this question however is that, although V and $V_\mathbb{R}$ contain the same elements, the concept of 'linear independence' has a different meaning in the two cases. In V, it means no non-trivial sum of the vectors vanishes, where the coefficients in the sum are *complex* numbers. In $V_\mathbb{R}$, however, only *real* numbers are allowed as expansion coefficients and hence two vectors such as $\mathbf{v}$ and $i\mathbf{v}$ ($i=\sqrt{-1}$) which are linearly dependent in V, are linearly *independent* in $V_\mathbb{R}$.

    In particular, if $\{\mathbf{e}^1, \mathbf{e}^2, \ldots, \mathbf{e}^n\}$ is a basis set for V, then the set of vectors $\{\mathbf{e}^1, i\mathbf{e}^1, \mathbf{e}^2, i\mathbf{e}^2, \ldots, \mathbf{e}^n, i\mathbf{e}^n\}$ is linearly independent in $V_\mathbb{R}$ and hence $\dim(V_\mathbb{R}) \geq 2n$. However, if $\mathbf{v}$ is a vector in $V_\mathbb{R}$ that cannot be expanded in terms of this set of vectors (with real coefficients) then it must also be impossible to expand $\mathbf{v}$ in V as a linear combination of $\{\mathbf{e}^1, \mathbf{e}^2, \ldots, \mathbf{e}^n\}$ with complex coefficients. But this contradicts the fact that this is a basis set for V, and hence there is no such $\mathbf{v} \in V_\mathbb{R}$. Thus the maximum number of linearly independent vectors in $V_\mathbb{R}$ is $2n$ and hence $\dim(V_\mathbb{R}) = 2n$.

    (b) Every vector $(a_1, a_2, \ldots, a_n)$ in $\mathbb{C}^n$ can be written in the form $(\mathrm{Re}(a_1) + i\,\mathrm{Im}(a_1), \mathrm{Re}(a_2) + i\,\mathrm{Im}(a_2), \ldots, \mathrm{Re}(a_n) + i\,\mathrm{Im}(a_n))$ which suggests at once that a suitable map $j : \mathbb{C}_\mathbb{R}^n \to \mathbb{R}^{2n}$ might be

$$j(a_1, a_2, \ldots, a_n) := (\mathrm{Re}(a_1), \mathrm{Im}(a_1), \ldots, \mathrm{Re}(a_n), \mathrm{Im}(a_n)).$$

This map is clearly a bijection, and it is easy to check that it is a *real* linear map between these two real vector spaces. And hence it is an isomorphism.

(c) In order that a subset W of a vector space V be a linear subspace of V, it is necessary that a sum of any two vectors in W should also lie in W, and that μw should lie in W if $w \in W$ and if μ is any complex (respectively real) number if V is a complex (respectively real) vector space. Evidently, the conditions are less restrictive in the real case and hence it is possible for a subset W of a complex vector space V to fail to be a linear subspace of V and yet to be a linear subspace of the real vector space $V_{\mathbb{R}}$.

This is precisely what happens in the case of the set H of all $n \times n$ hermitian matrices: $H := \{A$ in $M(n, \mathbb{C})$ such that $A^\dagger = A\}$.

If $A, B \in H$ then

$$(A+B)^\dagger = A^\dagger + B^\dagger = A + B$$

and hence $A + B \in H$. However, the adjoint of μA, where μ is a complex number, is $(\mu A)^\dagger = \mu^* A$ and hence μA is hermitian (i.e., lies in the subset H) only if μ is a real number. Which is precisely the statement that H is a linear subspace of the real vector space $M(n, \mathbb{C})_{\mathbb{R}}$ but not of the complex vector space $M(n, \mathbb{C})$.

29. (a) A vector space structure on the dual V^* of V can be defined in a way analogous to that in which the set of all linear operators on V is made into a vector space [cf. Eq.(2.4.2)].

If $F_1, F_2 \in V^*$ then $(F_1 + F_2)(v) := F_1(v) + F_2(v)$ for all $v \in V$.

If $F \in V^*$ and $\mu \in \mathbb{C}$ then $(\mu F)(v) := \mu F(v)$ for all $v \in V$.

It is a matter of trivial, and boring, detail to show that this does indeed define a complex vector space structure on V^*.

(b) If $\{e^1, e^2, \ldots, e^n\}$ is an orthonormal basis set for $\mathscr{H}$ then any $v \in \mathscr{H}$ can be expanded as

$$v = \sum_{i=1}^{n} v_i e^i, \qquad \text{where } v_i := \langle e^i, v \rangle \tag{1}$$

and hence, since $F : \mathscr{H} \to \mathbb{C}$ is a linear map,

$$F(v) = \sum_{i=1}^{n} v_i F(e^i). \tag{2}$$

On the other hand, if $w \in \mathscr{H}$ then $\langle w, v \rangle = \sum_{i=1}^{n} w_i^* v_i$, and hence, if we define

$$w_F := \sum_{i=1}^{n} F(e^i)^* e^i \tag{3}$$

we will get $F(v) = \langle w_F, v \rangle$ for all $v \in \mathscr{H}$ as required.

Note that such a vector must be unique since, in general, if $\langle w, v \rangle = \langle w', v \rangle$ for all $v \in \mathscr{H}$ then $\langle w - w', v \rangle = 0$ for all v in $\mathscr{H}$ and in particular for all of the basis elements e^i, $i = 1$, $\ldots$, n. But this means that $w - w'$ has zero expansion coefficients with respect to this basis set, and hence $w - w' = 0$.

Note that if $F_1(e^i) = F_2(e^i)$ for all $i = 1, \ldots, n$ then $F_1(v) = F_2(v)$ for all $v \in \mathscr{H}$, i.e., $F_1 = F_2$. It follows that the map $F \leadsto w_F$ is an injective map from $\mathscr{H}^*$ into $\mathscr{H}$. However, if w is *any* vector in $\mathscr{H}$ then the map $v \leadsto \langle w, v \rangle$

is a linear function on $\mathcal{H}$ and hence a member of $\mathcal{H}^*$. Thus, the map $F \rightsquigarrow \mathbf{w}_F$ is actually a bijection between $\mathcal{H}^*$ and $\mathcal{H}$.

However,

$$\mathbf{w}_{(\mu_1 F_1 + \mu_2 F_2)} = \mu_1^* \mathbf{w}_{F_1} + \mu_2^* \mathbf{w}_{F_2} \tag{4}$$

and hence this bijection is not a linear map but instead what is known as an *anti-linear* map (cf the definition of an anti-linear operator in Sec. 2.6). However, this suggests that we can define a linear bijection (and hence an isomorphism) $j : \mathcal{H}^* \to \mathcal{H}$ by

$$j(F) := \sum_{i=1}^{n} F(\mathbf{e}^i) \, \mathbf{e}^i \tag{5}$$

and the obvious slight modifications to the arguments above show that this is indeed so. Hence, $\mathcal{H}^* \cong \mathcal{H}$.

Note. This theorem is *not* true in the infinite-dimensional case since the step leading to Eq. (2) involves taking F 'through' the sum and this can be justified by linearity only in the finite-dimensional case. In order to justify it when dim $\mathcal{H} = \infty$, it is necessary to restrict F to belong to the *continuous* functions. Otherwise, the dual is much bigger than $\mathcal{H}$ itself.

30. Let us suppose that we have a sequence of vectors $\mathbf{v}^1, \mathbf{v}^2, \ldots$ that converges strongly to two different vectors $\mathbf{v}$ and $\mathbf{v}'$ in V. Thus

$$\lim_{n \to \infty} \left\| \mathbf{v}^n - \mathbf{v} \right\| = 0 \quad \text{and} \quad \lim_{n \to \infty} \left\| \mathbf{v}^n - \mathbf{v}' \right\| = 0. \tag{1}$$

Then, for all positive integers n,

$$0 \le \|\mathbf{v}-\mathbf{v}'\| = \|(\mathbf{v}-\mathbf{v}^n) + (\mathbf{v}^n-\mathbf{v}')\| \le \|\mathbf{v}-\mathbf{v}^n\| + \|\mathbf{v}'-\mathbf{v}^n\|, (2)$$

where we have used the basic triangle inequality in the definition of a norm. Taking the limit $n \to \infty$ on the right hand side, we get $0 + 0$ since the sequence $\mathbf{v}^n$ converges strongly to both $\mathbf{v}$ and $\mathbf{v}'$. Thus $\|\mathbf{v}-\mathbf{v}'\| = 0$ which implies that $\mathbf{v}=\mathbf{v}'$ since $\|\mathbf{w}\| = 0$ implies that $\mathbf{w} = 0$.

31. (a) Let $\mathbf{S}^N := \displaystyle\sum_{i=1}^{N} \mu_i \mathbf{e}^i$. Then, since $\mathscr{H}$ is a complete space, the sequence of vectors $\mathbf{S}^N$ will converge strongly if and only if $\mathbf{S}^N$ is a Cauchy sequence. But, if $N \ge M$,

$$\|\mathbf{S}^N - \mathbf{S}^M\|^2 = \left\| \sum_{i=M}^{N} \mu_i \mathbf{e}^i \right\|^2 = \left\langle \sum_{i=M}^{N} \mu_i \mathbf{e}^i, \sum_{j=M}^{N} \mu_j \mathbf{e}^j \right\rangle = \left| \sum_{i=M}^{N} \mu_i \right|^2$$

and hence $\mathbf{S}^N$ is a Cauchy sequence of vectors in $\mathscr{H}$ if and only if $\displaystyle\sum_{i=1}^{N} \mu_i$ is a Cauchy sequence of complex numbers. But this is true if and only if this sequence converges, i.e.,

$$\sum_{i=1}^{\infty} |\mu_i|^2 < \infty.$$

QED.

(b) We define a map $j: \ell_2 \to \mathscr{H}$ by

$$j(a_1, a_2, \ldots) := \sum_{i=1}^{\infty} a_i \mathbf{e}^i \qquad (1)$$

which is well-defined since the sequence $(a_1, a_2, \ldots)$ of complex numbers belongs to ℓ_2 if and only if $\displaystyle\sum_{i=1}^{\infty} |a_i|^2 < \infty$ and, according to the above, this is also the condition for the sum of vectors on the right hand side of Eq. (1) to belong to $\mathcal{H}$. This same remark also shows that $j : \ell_2 \to \mathcal{H}$ is surjective and, since it is clearly injective and linear, it follows that j is an isomorphism between ℓ_2 and $\mathcal{H}$.

32. Since F is a continuous linear function on $\mathcal{H}$, we can procede in analogy with the proof in Q. 29 in which the theorem we are seeking was proved for the case when $\mathcal{H}$ was finite-dimensional.

(**Note.** Every linear function on a finite-dimensional Hilbert space is continuous. Exercise.) Thus we take an orthonormal basis set $\{\mathbf{e}^1, \mathbf{e}^2, \ldots\}$ for $\mathcal{H}$ and expand any $\mathbf{v} \in \mathcal{H}$ as

$$\mathbf{v} = \sum_{i=1}^{\infty} v_i \mathbf{e}^i, \quad \text{where } v_i := \langle \mathbf{e}^i, \mathbf{v} \rangle. \tag{1}$$

Then, since F is continuous,

$$F(\mathbf{v}) = F\left(\lim_{N \to \infty} \sum_{i=1}^{N} v_i \mathbf{e}^i \right) = \lim_{N \to \infty} \left(\sum_{i=1}^{N} v_i F(\mathbf{e}^i) \right). \tag{2}$$

Now, define

$$\mathbf{w}^N := \sum_{i=1}^{N} F(\mathbf{e}^i) * \mathbf{e}^i. \tag{3}$$

Then,

$$\langle \mathbf{w}^N, \mathbf{v} \rangle = \sum_{i=1}^{N} v_i F(\mathbf{e}^i) \tag{4}$$

and hence, *if* the limit $N \to \infty$ exists for the sequence $\mathbf{w}^N$, we can take this as the desired vector $\mathbf{w}_F$ since, for all $\mathbf{v} \in \mathcal{H}$,

$$\langle \mathbf{w}_F, \mathbf{v} \rangle = \langle \lim_{N \to \infty} \mathbf{w}^N, \mathbf{v} \rangle = \lim_{N \to \infty} \langle \mathbf{w}^N, \mathbf{v} \rangle$$

$$= \lim_{N \to \infty} \sum_{i=1}^{N} v_i F(\mathbf{e}^i) = F(\mathbf{v})$$

as required.

Hence the main task is to show that $\mathbf{w}^N$ is a strongly convergent sequence of vectors in $\mathcal{H}$. According to the result in Q. 31(a), this will be true if and only if

$$\sum_{i=1}^{\infty} |F(\mathbf{e}^i)|^2 < \infty. \tag{5}$$

But,

$$\left\| \mathbf{w}^N \right\|^2 = \sum_{i=1}^{N} |F(\mathbf{e}^i)|^2 = \sum_{i=1}^{N} F(\mathbf{e}^i) * F(\mathbf{e}^i) = F(\mathbf{w}^N) \tag{6}$$

and hence our task is to show that $\lim_{N \to \infty} \left\| \mathbf{w}^N \right\| < \infty$.

Suppose then that $\lim_{N \to \infty} \left\| \mathbf{w}^N \right\| = \infty$. Then define $\mathbf{u}^N :=$ $\mathbf{w}^N / \left\| \mathbf{w}^N \right\|^2$ for which, from Eq. (6),

$$F(\mathbf{u}^N) = F(\mathbf{w}^N) / \left\| \mathbf{w}^N \right\|^2 = 1, \Rightarrow \lim_{N \to \infty} F(\mathbf{u}^N) = 1. \tag{7}$$

On the other hand, $\left\| \mathbf{u}^N \right\| = 1 / \left\| \mathbf{w}^N \right\|$ and so $\lim_{N \to \infty} \left\| \mathbf{u}^N \right\| = 0$ which implies that $\mathbf{u}^N$ converges strongly to 0. But then, since F is continuous,

$$\lim_{N \to \infty} F(\mathbf{u}^N) = F\left(\lim_{N \to \infty} \mathbf{u}^N \right) = F(\mathbf{0}) = 0 \tag{8}$$

which contradicts Eq. (7). Hence, the original assumption that $\lim_{N \to \infty} \|\mathbf{w}^N\| = \infty$ must be false. Thus the limit is finite and hence Eq. (5) is true which in turn implies that $\mathbf{w}^N$ is a strongly convergent sequence of vectors. And hence the result.

QED.

33. (a) To show that $A^\dagger$ is bounded, we must look at $\|A^\dagger \mathbf{v}\|$ and show that it is bounded above by some constant multiple of $\|\mathbf{v}\|$. Thus,

$$\|A^\dagger \mathbf{v}\|^2 = \langle A^\dagger \mathbf{v}, A^\dagger \mathbf{v} \rangle = \langle \mathbf{v}, A A^\dagger \mathbf{v} \rangle$$

$$\leq \|\mathbf{v}\| \, \|A A^\dagger \mathbf{v}\| \quad \text{(by Schwarz inequality)}$$
$$\leq \|\mathbf{v}\| \, \|A\| \|A^\dagger \mathbf{v}\| \quad \text{(since } A \text{ is a bounded operator).}$$

Thus,

$$\|A^\dagger \mathbf{v}\| \leq \|\mathbf{v}\| \, \|A\| \qquad \text{for all } \mathbf{v} \in \mathscr{H} \tag{1}$$

which shows that $A^\dagger$ is bounded.
Since $\|A^\dagger\|$ is the smallest number b such that $\|A^\dagger \mathbf{v}\| \leq b \|\mathbf{v}\|$ for all $\mathbf{v} \in \mathscr{H}$, Eq. (1) shows that

$$\|A^\dagger\| \leq \|A\|. \tag{2}$$

However, we also know that $(A^\dagger)^\dagger = A$ and hence, repeating the argument above with $A^\dagger$ and A interchanged, we find

$$\|A\| \leq \|A^\dagger\| \tag{3}$$

and so, from Eqs. (3–4), we conclude that $\|A^\dagger\| = \|A\|$.

(b) We must look at $\|AB\mathbf{v}\|$ for all $\mathbf{v}$ in $\mathcal{H}$. Thus,

$$\|AB\mathbf{v}\|^2 = \langle AB\mathbf{v}, AB\mathbf{v} \rangle = \langle B\mathbf{v}, A^\dagger AB\mathbf{v} \rangle$$

$$\leq \|B\mathbf{v}\|\|A^\dagger AB\mathbf{v}\| \quad \text{(by Schwarz inequality)}$$

$$\leq \|B\|\|\mathbf{v}\|\|A^\dagger\|\|AB\mathbf{v}\| \text{ (since } A^\dagger \text{ and } B \text{ are bounded).}$$

Thus,

$$\|AB\mathbf{v}\| \leq \|A\|\|B\|\|\mathbf{v}\| \quad \text{for all } \mathbf{v} \in \mathcal{H}$$

which shows that AB is bounded and that $\|AB\| \leq \|A\|\|B\|$.

34. (a) We have $(\mathbb{1} - P)^\dagger = \mathbb{1}^\dagger - P^\dagger = \mathbb{1} - P$ and hence $(\mathbb{1} - P)$ is hermitian. And, $(\mathbb{1} - P)^2 = (\mathbb{1} - P)(\mathbb{1} - P) = \mathbb{1} - 2P + P^2 = \mathbb{1} - P$.

Furthermore, for all $\mathbf{v} \in \mathcal{H}$,

$$\|(\mathbb{1}-P)\mathbf{v}\|^2 = \langle (\mathbb{1}-P)\mathbf{v}, (\mathbb{1}-P)\mathbf{v} \rangle = \langle \mathbf{v}, (\mathbb{1}-P)(\mathbb{1}-P)\mathbf{v} \rangle$$

$$= \langle \mathbf{v}, (\mathbb{1}-P)\mathbf{v} \rangle \leq \|\mathbf{v}\|\|(\mathbb{1}-P)\|.$$

Thus $(\mathbb{1}-P)$ is a bounded, hermitian operator, with $(\mathbb{1} - P)^2 = (\mathbb{1} - P)$ and hence it is a projection operator.

(b) Using the same type of argument that we have employed several times already, we have

$$\|P\mathbf{v}\|^2 = \langle P\mathbf{v}, P\mathbf{v} \rangle = \langle \mathbf{v}, P^2\mathbf{v} \rangle = \langle \mathbf{v}, P\mathbf{v} \rangle$$

$$\leq \|\mathbf{v}\|\|P\mathbf{v}\| \quad \text{for all } \mathbf{v} \in \mathcal{H}.$$

Thus $\|P\mathbf{v}\| \leq \|\mathbf{v}\|$, which shows that $\|P\| \leq 1$.

(c) Suppose that $P\mathbf{v} = \mu\mathbf{v}$ so that μ is an eigenvalue of P with eigenvector $\mathbf{v}$. Then

$$\mu\mathbf{v} = P\mathbf{v} = P^2\mathbf{v} = P(P\mathbf{v}) = P(\mu\mathbf{v}) = \mu^2\mathbf{v}$$

and hence, since $\mathbf{v} \neq 0$, we have $\mu - \mu^2 = 0$, i.e., $\mu(1-\mu) = 0$.

Thus the two possible eigenvalues for P are $\mu = 1$ and $\mu = 0$.

Both do actually occur since if $\mathbf{v}$ is any vector in $\mathcal{H}$ such that $P\mathbf{v} \neq 0$ (and there must be at least one or else P would be the 0 operator) then $P(P\mathbf{v}) = P^2\mathbf{v} = P\mathbf{v}$ so that $P\mathbf{v}$ is an eigenvector of P with eigenvalue $\mu = 1$.

Similarly, $P(\mathbb{1}-P)\mathbf{v} = (P - P^2)\mathbf{v} = \mathbf{0}$ for all $\mathbf{v} \in \mathcal{H}$ and hence any vector $\mathbf{v}$ such that $(\mathbb{1} - P)\mathbf{v} \neq 0$ will give a non-trivial eigenvector $(\mathbb{1}-P)\mathbf{v}$ with eigenvalue 0. There will always be such a vector $\mathbf{v}$ provided $P \neq \mathbb{1}$.

We have already shown that $\|P\| \leq 1$. But if $\mathbf{v}$ is an eigenvector with eigenvalue 1 then $P\mathbf{v}=\mathbf{v}$ and this implies that $\|P\| \geq 1$. Hence $\|P\| = 1$.

(d) To prove hermiticity, we must show that $\langle \mathbf{u}, P\mathbf{v} \rangle = \langle P\mathbf{u}, \mathbf{v} \rangle$ for all $\mathbf{u}, \mathbf{v}$ in $\mathcal{H}$. But $P\mathbf{v} = \mathbf{v}_W$ and therefore

$$\langle \mathbf{u}, P\mathbf{v} \rangle = \langle (\mathbf{u}_W + \mathbf{u}_{W_\perp}), \mathbf{v}_W \rangle = \langle \mathbf{u}_W, \mathbf{v}_W \rangle, \qquad (1)$$

where we have used the unique decomposition $\mathbf{u} = \mathbf{u}_W + \mathbf{u}_{W_\perp}$ with $\mathbf{u}_W \in W$ and $\mathbf{u}_{W_\perp} \in W_\perp$. Similarly,

$$\langle P\mathbf{u}, \mathbf{v} \rangle = \langle \mathbf{u}_W, (\mathbf{v}_W + \mathbf{v}_{W_\perp}) \rangle = \langle \mathbf{u}_W, \mathbf{v}_W \rangle \qquad (2)$$

and so comparing Eqs. (1) and (2), we see that $\langle P\mathbf{u}, \mathbf{v} \rangle = \langle \mathbf{u}, P\mathbf{v} \rangle$ for all $\mathbf{u}, \mathbf{v} \in \mathcal{H}$. Thus P is a hermitian operator.

If $\mathbf{w}$ is a vector in W then $P\mathbf{w} = \mathbf{w}$ and so, in particular, $P(P\mathbf{v}) = P\mathbf{v}$ for all $\mathbf{v} \in \mathcal{H}$. But this means precisely that $P^2 = P$. We have already shown in part (a) of the question that any hermitian operator Q satisfying $Q^2 = Q$ is bounded [we proved it for $Q = (\mathbb{1} - P)$ but the method applies at once to the general case] and hence P is a projection operator.

Note. It is clear that

$$W = \{\mathbf{v} \text{ in } \mathscr{H} \text{ such that } P\mathbf{v} = \mathbf{v}\} \qquad (3)$$
$$W_\perp = \{\mathbf{v} \text{ in } \mathscr{H} \text{ such that } P\mathbf{v} = \mathbf{0}\}. \qquad (4)$$

Thus, W and $W_\perp$ are the eigenspaces corresponding to the two eigenvalues 1 and 0, respectively of the operator P. It is noteworthy that the converse situation also applies, i.e., *every* projection operator on $\mathscr{H}$ can be associated with a particular closed subspace W such that $\mathscr{H} \cong W \oplus W_\perp$. One simply *defines* W by Eq. (3), i.e., it is the set of all eigenvectors with eigenvalue 1, and then it can be shown that W is a closed linear subspace of $\mathscr{H}$ and that $W_\perp$ is precisely the set of all eigenvectors with eigenvalue 0.

The conclusion is that there is a one-to-one correspondence between closed linear subspaces of $\mathscr{H}$ and projection operators. This tying together of a geometrical picture with the more algebraic concept of a hermitian operator satisfying $P^2 = P$ is very valuable and it is useful to be able to switch from one to the other as the need dictates.

35. The matrix representatives of A with respect to the basis sets $\{\mathbf{e}^1, \mathbf{e}^2, \ldots, \mathbf{e}^n\}$ and $\{\mathbf{f}^1, \mathbf{f}^2, \ldots, \mathbf{f}^n\}$ are the $n \times n$ matrices M and N defined respectively by

$$A\mathbf{e}^i = \sum_{j=1}^n \mathbf{e}^j M_{ji}, \quad A\mathbf{f}^i = \sum_{j=1}^n \mathbf{f}^j N_{ji}. \qquad (1)$$

Since each member $\mathbf{e}^i$ of the first basis set can be expanded in terms of the elements of the second set, and vice versa, there must be some $n \times n$, invertible matrix $B \in GL(n, \mathbb{C})$ such that

$$\mathbf{f}^i = \sum_{k=1}^{n} \mathbf{e}^k B_{ki}, \quad i = 1, \ldots, n. \tag{2}$$

Operating on both sides of this equation, and using Eq. (1), we get

$$A\mathbf{f}^i = \sum_{k=1}^{n} (A\mathbf{e}^k) B_{ki} = \sum_{k=1}^{n} \sum_{j=1}^{n} \mathbf{e}^j M_{jk} B_{ki}. \tag{3}$$

But the left hand side of this expression can be rewritten as

$$A\mathbf{f}^i = \sum_{k=1}^{n} \mathbf{f}^k N_{ki} = \sum_{k=1}^{n} \sum_{j=1}^{n} \mathbf{e}^j B_{jk} N_{ki} \tag{4}$$

and comparing Eqs. (3) and (4) [and remembering that $\{\mathbf{e}^1, \mathbf{e}^2, \ldots, \mathbf{e}^n\}$ is a set of linearly independent vectors], we get

$$\sum_{k=1}^{n} B_{jk} N_{ki} = \sum_{k=1}^{n} M_{jk} B_{ki} \tag{5}$$

or, in matrix form, $BN = MB$ and hence, since B is invertible, we find that the matrix representatives of the two basis sets are related by

$$M = BNB^{-1}$$

as claimed.

36. Using the results of Q. 14, we write the group S_3 as the set of six elements $\{e, x, y, y^2, xy, xy^2\}$ with

$$x^2 = y^3 = e, \quad yx = xy^2, \quad y^2x = xy, \tag{1}$$

where

$e := 123, \; x := 213, \; y := 321, \; y^2 = 231, \; xy = 321, \; xy^2 = 132.$

Since T is a one-dimensional complex representation, the vector space is simply $\mathbb{C}$ and each $T(g)$, $g \in S_3$, is an invertible complex number that acts on a vector $\mathbf{v} \in \mathbb{C}$ (i.e., any complex number) by the usual multiplication of complex numbers.

Then $T(e) = 1$ and $x^2 = e$ implies $[T(x)]^2 = 1$, hence

$$T(x) = +1 \text{ or } -1. \tag{2}$$

Similarly, the relation $y^3 = e$ implies $[T(y)]^3 = 1$ and hence

$$T(y) = 1, \text{ or } e^{2i\pi/3}, \text{ or } e^{-2i\pi/3}. \tag{3}$$

However, the relation $y^2x = xy$ implies that $[T(y)]^2 \, T(x) = T(x) \, T(y)$ which, since $T(x) = \pm 1$, implies that $[T(y)]^2 = T(y)$, i.e., $T(y) = 1$ is the only choice from the three possibilities in Eq. (3) that is consistent with this relation.

Thus $T(x) = \pm 1$ and $T(y) = 1$, which is clearly compatible with the remaining relation $yx = xy^2$. A consistent representation can now be obtained by simply *defining* the remaining operators by their algebraic expressions in $T(x)$ and $T(y)$; for example, $T(xy) := T(x) \, T(y)$.

Thus we see that there are two one-dimensional complex representations of the group S_3:

(a) The trivial representation with $T(g) = 1$ for all $g \in S_3$

(b) $T(e) = T(y) = T(y^2) = 1$

$T(x) = T(xy) = T(xy^2) = -1.$ \hspace{2cm} (4)

The kernel of the trivial representation is of course the entire group S_3.

The kernel of the non-trivial representation in Eq. (4) is the subgroup of S_3 made up of the elements $\{e, y, y^2\}$ which, since $y^3 = e$, is a $\mathbb{Z}_3$ group.

37. Since every representation of a finite group is equivalent to a unitary representation. there is no loss of generality in assuming that the two-dimensional representation that we are seeking is in terms of 2×2 unitary matrices on the vector space $\mathbb{C}^2$.

Furthermore, since we are only interested in the representation up to similarity transformations, it is possible to demand that any one particular group element be represented by a diagonal matrix. (This corresponds to taking a basis set for $\mathbb{C}^2$ made up of the eigenvectors of that particular operator.) So, we let $U(y) = \begin{pmatrix} p & 0 \\ 0 & q \end{pmatrix}$ be the unitary matrix representing $y \in S_3$. The unitary condition requires that $|p|^2 = |q|^2 = 1$, and the condition $y^3 = e$ implies that $[U(y)]^3 = 1$ so that both p and q are cube-roots of unity, i.e.,

$$p^3 = q^3 = 1. \hspace{2cm} (1)$$

We cannot assume that $U(x)$ is a diagonal matrix since $xy \neq yx$ implies that $U(x) U(y)$ may not equal $U(y) U(x)$, and operators need to commute in order to be simultaneously diagonalizable. Thus we write $U(x) = \begin{pmatrix} a & b \\ c & d \end{pmatrix}$ and we must work out the constraints on the complex numbers a, b, c, d, p, q in order that the remaining constraints $x^2 = e$, $yx = xy^2$ and $y^2x = xy$ be satisfied.

There are a number of ways in which one can uncover the implications of these constraints. An efficient start is made by noticing that the equation

$$U(y)\,U(x) = U(x)\,[U(y)]^2 \qquad (2)$$

implies that $\det\,[U(y)] = 1$ so that $pq = 1$ or, since $p^3 = 1$, $q = p^2$.
Thus

$$U(y) = \begin{pmatrix} p & 0 \\ 0 & p^2 \end{pmatrix} \quad \text{with } p^3 = 1. \qquad (3)$$

The relations $yx = xy^2$ and $y^2x = xy$ imply respectively $U(y)\,U(x) = U(x)[U(y)]^2$ and $[U(y)]^2\,U(x) = U(x)\,U(y)$ which, using Eq. (3), read

$$\begin{pmatrix} pa & pb \\ p^2c & p^2d \end{pmatrix} = \begin{pmatrix} ap^2 & bp \\ cp^2 & dp \end{pmatrix} \quad \text{and} \quad \begin{pmatrix} p^2a & p^2b \\ pc & pd \end{pmatrix} = \begin{pmatrix} ap^2 & bp^2 \\ cp & dp \end{pmatrix} \qquad (4)$$

which implies that $ap(1 - p) = dp(1 - p) = 0$ or, since $p \neq 0$ ($p^3 = 1$),

$$a(1 - p) = d(1 - p) = 0. \qquad (5)$$

One solution to Eq. (5) is $p = 1$. However, this would imply that $U(y) = \mathbb{1}$ which would therefore commute with $U(x)$. This means that $U(x)$ could be diagonalized simultaneously with $U(y)$, and the rest of the matrices $U(g)$, $g \in S_3$, would also be diagonal as they can be written as simple products of $U(x)$ and $U(y)$. But then the representation would not be irreducible as the column vectors in $\mathbb{C}^2$ of the form $\begin{pmatrix} \mu \\ 0 \end{pmatrix}$ would be a S_3-invariant subspace of $\mathbb{C}^2$ [and ditto for $\begin{pmatrix} 0 \\ \mu \end{pmatrix}$].

Thus we must choose the other solution to Eq. (5), namely $a = d = 0$. But then the constraint $x^2 = e$ implies $[U(x)]^2 = \mathbb{1}$ and hence $bc = 1$. Furthermore, unitarity of $U(x)$ gives $|b|^2 = |c|^2 = 1$. And so, summarizing so far we have

$$U(x) = \begin{pmatrix} 0 & b \\ b^{-1} & 0 \end{pmatrix} |b| = 1; \quad U(y) = \begin{pmatrix} p & 0 \\ 0 & p^2 \end{pmatrix} p^3 = 1 \quad (6)$$

and the task now is to see to what extent the possible different values for p and b correspond to different similarity transformations.

Spotting what to do quickly is partly a question of familiarity with the manipulation of 2×2 matrices. For our purposes, the following result is useful

$$\begin{pmatrix} z & 0 \\ 0 & 1/z \end{pmatrix} \begin{pmatrix} a & b \\ c & d \end{pmatrix} \begin{pmatrix} 1/z & 0 \\ 0 & z \end{pmatrix} = \begin{pmatrix} a & bz^2 \\ c/z^2 & d \end{pmatrix} \quad (7)$$

which is a similarity transformation with a unitary matrix if $|z| = 1$. Thus, if we choose z such that $z^2 = 1/b$, we get a similarity transformation to the unitary representation

$$U(x) = \begin{pmatrix} 0 & 1 \\ 1 & 0 \end{pmatrix}, \quad U(y) = \begin{pmatrix} p & 0 \\ 0 & p^2 \end{pmatrix} \quad (8)$$

with $p^3 = 1$ (but not with $p = 1$). There are just two possibilities, namely, $p = e^{+2i\pi/3}$ and $p = e^{-2i\pi/3}$ corresponding to the two matrices:

$$U(y) = \begin{pmatrix} e^{+2i\pi/3} & 0 \\ 0 & e^{-2i\pi/3} \end{pmatrix} \text{ or } \begin{pmatrix} e^{-2i\pi/3} & 0 \\ 0 & e^{+2i\pi/3} \end{pmatrix}.$$

However, these can be interchanged with a similarity transformation with the matrix $\begin{pmatrix} 0 & 1 \\ 1 & 0 \end{pmatrix}$ which leaves $U(x)$ alone.

Thus, up to similarity transformations, there is a unique irreducible two-dimensional representation of the group S_3

$$p := e^{2i\pi/3},$$

$$U(e) = \begin{pmatrix} 1 & 0 \\ 0 & 1 \end{pmatrix}, \ U(x) = \begin{pmatrix} 0 & 1 \\ 1 & 0 \end{pmatrix}, \ U(y) = \begin{pmatrix} p & 0 \\ 0 & p^2 \end{pmatrix},$$

$$U(y^2) = \begin{pmatrix} p^2 & 0 \\ 0 & p \end{pmatrix}, \ U(xy) = \begin{pmatrix} 0 & p^2 \\ p & 0 \end{pmatrix}, \ U(xy^2) = \begin{pmatrix} 0 & p \\ p^2 & 0 \end{pmatrix}. \tag{9}$$

Clearly this representation is faithful with trivial kernel.

38. (a) We must show that $T(g_1)\,T(g_2) = T(g_1\,g_2)$ for all $g_1, g_2 \in G$.
But,

$$(T(g_1)\,T(g_2)f)\,(x) = (T(g_1)\,(T(g_2)f))\,(x)$$
$$= (T(g_2)f)\,(g_1^{-1}x)$$
$$= f(g_2^{-1}\,g_1^{-1}x)$$
$$= f((g_1\,g_2)^{-1}\,x)$$
$$= (T(g_1\,g_2)f)\,(x)$$

as required.

(b) We can check unitarity directly

$$\langle T(g)f_2, T(g)f_1 \rangle = \sum_{x \in X} f_2^*(g^{-1}x)f_1(g^{-1}x). \qquad (1)$$

But to say that G acts on X means that each $g \in G$ acts as a bijection of X, and therefore as the point $x \in X$ ranges once over each member of X in the sum in Eq. (1), the point gx (with g fixed) also ranges once over each member of X. Thus the right hand side of (1) is equal to

$$\sum_{x \in X} f_2^*(x)f_1(x) = \langle f_2, f_1 \rangle$$

and hence the representation is unitary.

39. (a) In general, the matrix version of a linear operator A with respect to any basis set in the Hilbert space is defined by

$$Ae^i = \sum_{j=1}^{n} e^j A_{ji} \quad i = 1, \ldots, n = \dim \mathcal{H} < \infty. \qquad (1)$$

If the basis set $\langle e^1, e^2, \ldots, e^n \rangle$ is orthonormal then taking the scalar product of both side of Eq. (1) with a particular element e^k gives,

$$\langle e^k, Ae^i \rangle = A_{ki} \quad k, i = 1, \ldots, n \qquad (2)$$

as an explicit expression for the matrix A_{ij} representing the operator A.

Thus, in the case of the regular representation, the matrix $M(g)$ representing the operator $R(g)$ with respect to the orthonormal basis vectors $\{f^g, g \in G\}$ is a $|G| \times |G|$ matrix, whose matrix elements $M_{g_1 g_2}$ are labelled by the pair g_1 (the row index) and g_2 (the column index), defined by

$$M_{g_1 g_2}(g) := \langle \mathbf{f}^{g_1}, R(g) \, \mathbf{f}^{g_1} \rangle = \sum_{a \in G} \mathbf{f}^{g_1 *}(a) \, \mathbf{f}^{g_2}(g^{-1} a)$$

$$= \mathbf{f}^{g_2}(g^{-1} g_1).$$

Thus the matrix representation of the left regular representation with respect to this particular orthonormal basis is

$$M_{g_1 g_2}(g) = 1 \text{ if } g_2 = g^{-1} g_1, \text{ i.e., if } g = g_1 g_2^{-1}$$

$$= 0 \text{ otherwise.} \tag{3}$$

(b) If $G = \mathbb{Z}_3 = \{e, a, a^2\}$ with $a^3 = e$, then the regular representation is three-dimensional since $|\mathbb{Z}_3| = 3$. Direct calculation then shows that the matrix version is

$$M(e) = \begin{pmatrix} 1 & 0 & 0 \\ 0 & 1 & 0 \\ 0 & 0 & 1 \end{pmatrix}, M(a) = \begin{pmatrix} 0 & 0 & 1 \\ 1 & 0 & 0 \\ 0 & 1 & 0 \end{pmatrix},$$

$$M(a^2) = \begin{pmatrix} 0 & 1 & 0 \\ 0 & 0 & 1 \\ 1 & 0 & 0 \end{pmatrix}.$$

(c) We must check that $S(g_1) S(g_2) = S(g_1 g_2)$. But, if $f \in \text{Map } (G, \mathbb{C})$,

$$(S(g_1) S(g_2) f) (g') = (S(g_1) (S(g_2) f)) (g')$$
$$= (S(g_2) f) (g' g_1)$$
$$= f(g' g_1 g_2) = (S(g_1 g_2) f) (g')$$

as required.

The presence or absence of g^{-1} factors in the left and right regular representations suggests that a possible intertwining operator is the map $A : \text{Map } (G, \mathbb{C}) \to \text{Map } (G, \mathbb{C})$ defined by

$$(Af) (g) := f(g^{-1}). \tag{4}$$

First we must check and see if it is a unitary operator. Thus,

$$\langle Af_1, Af_2 \rangle = \sum_{g \in G} (Af_1)^*(g) (Af_2) (g) = \sum_{g \in G} f_1^* (g^{-1}) f_2 (g^{-1})$$

$$= \sum_{g \in G} f_1^*(g) f_2 (g) \tag{5}$$

since summing over g is the same as summing over g^{-1}.

Hence $\langle Af_1, Af_2 \rangle = \langle f_1, f_2 \rangle$ for all $f_1, f_2 \in \text{Map}(G, \mathbb{C})$. Furthermore, $A^2 = \mathbb{1}$ and hence A^{-1} exists, and in fact is equal to A. Therefore A is a unitary operator.
Now we look at $AR(g) A^{-1}$

$$(AR(g) A^{-1} f)(g')$$
$$= (R(g) A^{-1} f) (g'^{-1}) = (A^{-1} f) (g^{-1} g'^{-1})$$

$$= (A^{-1}f)\,((g'\,g)^{-1}) = f(g'\,g)$$
$$= (S(g)f)\,(g')$$

which, since it is true for all $g' \in G$, implies that

$$AR(g)\,A^{-1} = S(g) \quad \text{for all } g \in G$$

and hence the left and right regular representations of the finite group G are unitarily equivalent.

40. The orthogonality relations of interest are contained in Eq.(3.5.6) of the lecture notes, and are of the form:

$$\langle \chi_\mu, \chi_{\mu'} \rangle = \delta_{\mu\mu'}, \tag{1}$$

where χ_μ and $\chi_{\mu'}$ are the characters of the irreducible representations of G labelled by μ and μ', respectively, and the correctly normalized scalar product on Map $(G, \mathbb{C})$ is

$$\langle f_1, f_2 \rangle := \frac{1}{|G|} \sum_{g \in G} f_1^*(g)\, f_2(g). \tag{2}$$

Let us denote the characters corresponding to the trivial and non-trivial one-dimensional representation of S_3 in Eq. (4) of Q. 36, by $\chi_{\mathbf{1}}(g)$ and $\chi_1(g)$, respectively and let $\chi_2(g)$ denote the character of the two-dimensional representation in Eq. (9) of Q. 37. Since $\chi(g) := \text{Tr } U(g)$, it is possible to read off the values of the characters in our case using the explicit matrix representations in Eqs. (4) and (9). [Of course, the character of a one-dimensional representation is just the complex number $U(g)$ itself.]

These results can be tabulated in the form (where $p = e^{2i\pi/3}$, which implies that $p+p^2 = -1$)

	χ_1	χ_1	χ_2
e	1	1	2
x	1	-1	0
y	1	1	$p+p^2 = -1$
y^2	1	1	$p^2+p = -1$
xy	1	-1	0
xy^2	1	-1	0

Thus,

$$\langle \chi_1, \chi_1 \rangle = \frac{1}{6}(1^2 + 1^2 + 1^2 + 1^2 + 1^2 + 1^2) = 1 \tag{3}$$

$$\langle \chi_1, \chi_1 \rangle = \frac{1}{6}(1^2 + (-1)^2 + 1^2 + 1^2 + (-1)^2 + (-1)^2) = 1 \tag{4}$$

$$\langle \chi_2, \chi_2 \rangle = \frac{1}{6}(2^2 + 0^2 + (-1)^2 + (-1)^2 + 0^2 + 0^2) = 1 \tag{5}$$

$$\langle \chi_1, \chi_2 \rangle = \frac{1}{6}(1 \times 2 + (-1) \times 0 + 1 \times (-1)$$
$$+ 1 \times (-1) + (-1) \times 0 + (-1) \times 0)$$
$$= \frac{1}{6}(2 - 1 - 1) = 0 \tag{6}$$

which are all in agreement with the results predicted from Eq. (1).

Note that in general, the character of the trivial representation $U(g) = \mathbb{1}$ for all $g \in G$, is always some multiple of 1 (independent of $g \in G$) and that therefore the inner product of this character with any other character χ is always just $\sum_{g \in G} \chi(g)$. We can see at once from the table therefore that $\langle \chi_{1}, \chi_1 \rangle = \langle \chi_{1}, \chi_2 \rangle = 0$, which completes the verification of Eq. (1) in this S_3 example.

INDEX